Borders, Mobilities and Migrations

Perspectives from the Mediterranean, 19-21st Century

P.I.E. Peter Lang

Bruxelles · Bern · Berlin · Frankfurt am Main · New York · Oxford · Wien

Lisa Anteby-Yemini, Virginie Baby-Collin,
Sylvie Mazzella, Stéphane Mourlane,
Cédric Parizot, Céline Regnard & Pierre Sintès (eds.)

Borders, Mobilities and Migrations

Perspectives from the Mediterranean, 19-21st Century

"Regional Integration and Social Cohesion"
No.13

This book is published in the framework of the MIMED research program of the Mediterranean House of Humanities (MMSH) with the financial participation of the Regional Council Provence-Alpes-Côte d'Azur and of the Aix-Marseille University. We thank Brigitte Marin, Director of the MMSH.

Éditions scientifiques internationales
Brussels, 2014
1 avenue Maurice, B-1050 Brussels, Belgium
www.peterlang.com; info@peterlang.com

ISSN 2030-8787
ISBN 978-2-87574-123-3
D/2014/5678/25

Library of Congress Cataloging-in-Publication Data
Borders, mobilities and migrations : perspectives from the Mediterranean, 19-21st century / Lisa Anteby-Yemini, Virginie Baby-Collin, Sylvie Mazzella, Stéphane Mourlane, Cédric Parizot, Céline Regnard & Pierre Sintes (eds.).
pages cm. — (Regional integration and social cohesion, ISSN 2030-8787 ; No. 13)
Includes bibliographical references and index.
ISBN 978-2-87574-123-3 (alk. paper)
1. Mediterranean Region—Emigration and immigration—History. 2. Mediterranean Region—Emigration and immigration—Government policy. 3. Mediterranean Region—Emigration and immigration—Social aspects. 4. Migration, Internal—Mediterranean Region. I. Anteby-Yemini, Lisa, author, editor of compilation.
JV7590.B674 2014 304.809182'2—dc23 2013045196

CIP also available from the British Library, GB.

Bibliographic information published by "Die Deutsche Nationalbibliothek"

"Die Deutsche Nationalbibliothek" lists this publication in the "Deutsche Nationalbibliografie"; detailed bibliographic data is available on the Internet at <http://dnb.d-nb.de>.

Contents

Introduction 11
Lisa Anteby-Yemini, Virginie Baby-Collin, Sylvie Mazzella, Stéphane Mourlane, Cédric Parizot, Céline Regnard & Pierre Sintès

PART I. MOBILITY AND SYSTEMS OF CONTROL

Introduction 21
Sylvie Mazzella, Céline Regnard

Illegal Migration of Italians from the Liberal Age to Freedom of Movement Within the European Community. Geographic Directions, Public Attitudes and the Strategies of Migrants 25
Sandro Rinauro

Building Lines between Nations. Border-making and Police Practices at the French-Italian Frontier, 1890-1914 41
Ilsen About

North African Migration under Surveillance. Between Home Country and Host Country 57
Sylvie Mazzella

Introduction of Procedures for Migration Controls in Greece. Case Study in the Aegean Islands 71
Laurence Pillant

PART II. MOBILITY AND THE BORDER ECONOMY

Introduction 89
Cédric Parizot

An Undocumented Economy of Control. Workers, Smugglers and Sate Authorities in Southern Israel/Palestine 93
Cédric Parizot

Border Economies.
A Nascent Migration Industry Around Lampedusa 113
Heidrun Friese

"Maid in Lebanon".
Lebanon and the World Domestic Service Market 131
Assaf Dahdah

The Palestinian Olive Industry and Israel's Separation Policy.
Trading at the Border (2002-2012) 149
Arnaud Garcette

PART III. BEYOND STATE BORDERS:
MIGRANTS' MOBILITY IN EVERYDAY LIFE

Introduction 171
Lisa Anteby-Yemini, Virginie Baby-Collin

Between Embodied Cosmopolitism and Sexual Humanitarianism.
The Fractal Mobilities and Subjectivities
of Migrants Working in the Sex Industry 175
Nick Mai

African Asylum Seekers in Israel.
New Political, Racial and Urban Borders.. 193
Lisa Anteby-Yemini

Rendering People Illegal and Invisible.
Illegal Migrants and Daily Life (United States, Spain) 215
Virginie Baby-Collin

PART IV. CIRCULATION, FRONTIER AND MEMORY OF MIGRATIONS

Introduction 235
Stéphane Mourlane, Pierre Sintès

Migration and Nation. The Situation of Italians in the Eastern
Mediterranean in the Early 19th Century.. 239
Fabrice Jesné

Algerian Jews. How to Make the Return Journey?........................ 251
Karima Dirèche

"Invisibility" and Memory. Italian Immigration in France During the Second Half of the 20th Century 267
Stéphane Mourlane, Céline Regnard

New Geographical Mobilities on the Edges of Greece.................... 287
Pierre Sintès

Postface.. 301
Evelyne Ritaine

Shorts Biographies of the Authors... 307

Introduction[1]

Lisa ANTEBY-YEMINI, Virginie BABY-COLLIN,
Sylvie MAZZELLA, Stéphane MOURLANE,
Cédric PARIZOT, Céline REGNARD & Pierre SINTÈS

Globalization, as a process of accelerated commercial exchanges between different types of universality, is a long-term phenomenon that is often seen as part of the 15th century's Age of Discovery (Grataloup, 2006; Boucheron, 2009). This was the period when 'world economies' first began to be organized structurally through trade which led to the launch of international commerce as a form of capitalism with the Mediterranean as one of the main marketplaces (Braudel, 1993a, 1993b). In the 19th century, colonization and new industrial revolutions provided the bases for greater and more concentrated movements of people and goods (Wallerstein, 2006). The economic and political upheavals of the 19th and 20th centuries (industrial revolutions, colonial empires, war, decolonization, etc.) were accompanied by European migration and mobility which in turn led to the migratory phenomenon we see today. Simultaneously with intra-European labour migration and north-south migration across the Mediterranean, the movement of people ensured the settlement of colonial empires, and later of decolonialized countries. War and conflicts have also caused large displacements of populations. Over the last half-century, globalization has led to an unprecedented acceleration of mobility and trade, encouraged as much by major technological revolutions in transportation and communications as by the arrival of global capitalism. Globalized migration (Simon, 2008), defined as an unprecedented process of diversification and intensification of population flows, has transformed not only relations with otherness – as incarnated by the Other – but also with space and has thoroughly reconfigured the territories that it has brought into contact (Sassen, 1991; Lévy, 1999; Urry, 2000; Beck, 2003; Baduel, 2008).

---

1 The contributors to this volume wish to thank Caroline Mackenzie for her translation (from French) and correction of the articles published here, and Simon Turner for his translation (from Italian) of the article by Sandro Rinauro.

At the same time, with the consolidation of nations during the 19th century, the paradigm of the nation-state led to the installation of frontiers and establishment of mechanisms for controlling territories and identifying individuals in Europe and colonial empires (Foucher, 1986; Noiriel, 1992; Schnapper, 2001, About and Denis 2010). Tensions caused by increased movements of persons and goods and the closure of frontiers are again part of our reality today, as can be seen in current control and surveillance systems, often considered as the outcome of post-modernity, including biometric identification systems. Paradoxically, acceleration of mobilities is paralleled by a proliferation of controls that restrict movement and are made acceptable by doxa on the 'risk' that migration represents (Beck, 2003). Contemporary controls of mobility have drastically concentrated the operation and manifestations of national frontiers and frontier lines (borders) at strategic points (airports, maritime straits, islands). Extending both without and within the limits of national territory, frontiers are now flexible and multiple, pixellized and networked (Cuttita, 2006; Bigo, Bocco & Piermay, 2010). Control mechanisms at frontiers increasingly focus on the status of migrants ('illegal', 'without papers', asylum seekers', clandestine workers', visa holders, 'legal' or legitimate, etc.) and thus make it possible to detach the frontier from the territory and attach it to the individual. Thus, to a certain extent, each person carries the frontier 'on the sole of their shoes' and sets it down at the place where they take up residence.

This collection challenges the paradoxical articulation between, on the one hand, generalized mobilities and, on the other, processes by which controls are reinforced along frontiers profoundly transformed in their manifestations and operations, as part of long-term dynamics from the 19th to the 21st century. The Mediterranean in which this reflection has been constructed is a rich laboratory for observing these tensions between circulation and closure, and their consequences for practices as well as for construction of identities for the principal actors: migrants.

In order to study this paradox, we must take into account the ways in which the players evolve. In Europe, the tightening of external frontiers plays an integral role in complex processes that include not only national tactics, but also regional Euro-Mediterranean economic strategies and issues relating to European integration (Ritaine, 2012). In mobilizing supra- and infra-national political and institutional actors, the installation of control systems can no longer be analyzed uniquely from the perspective of a single nation (Wolff, 2008). Nevertheless we must recognize that, even if the role of nation-states remains essential today, nations are no longer masters of their territories and frontiers (Brown, 2009). The construction of the European area for free circulation goes hand-in-hand with the process of internalization and externalization of frontiers. Although European citizens are able to travel freely throughout the EU, 'others' are obliged to submit to controls which have become increasingly rigorous because of the technology and resources allocated to this purpose and more widespread throughout the region. Checks on 'clandestine migrants' can be carried out anywhere, and even in the streets of major cities, proof of the existence of these forms of internalized frontiers, one of the most tangible manifestations being detention centres. Externalized frontiers are set up through the transfer of identity controls to southern and eastern Mediterranean countries, which allows the modification of migrant flows – a phenomenon seen in other regions around the world (Audebert and Robin, 2009). Given the changes occurring in the construction of the European community and in the acceleration of economic growth in Northern Mediterranean countries, certain emigration regions became attractive regions for immigration since the 1980s (Wihtol de Wenden, 2009) – notably Spain, Italy and Greece, today facing a serious economic crisis – and moves towards tightening and externalizing Europe's outer frontiers contributes to the consolidation of zones where migrants gather on the Southern Mediterranean coast and in North African and Middle Eastern countries. These countries are not only transit zones but also, increasingly, areas for long-term settlement for migrants, creating holding 'nets' in 'buffer countries' located at the gates of Fortress Europe (Bredeloup and Pliez, 2005; Bel Air, 2006). In conflict zones, the multiplication and growing sophistication of control systems used to control and filter the passage of people provides opportunities for new forms of mobility, the Israeli-Palestinian zone being the most obvious example (Latte Abdallah and Parizot, 2011). The frontiers of the globalized world are therefore transforming control territories in the Mediterranean, not only because new systems are being deployed, but also because they must be constantly readjusted in order to take into account changes in crossing practices and the reorientation of flows adopted by migrant groups. The methods and practices used by migrants for crossing or avoiding frontiers have become more complex with each change in control systems (movement of frontiers, creation of inland detention centres, etc.) and with the evolution of crossings and passages in line with the greater sophistication of control systems (biometric, profiling, etc.) (Bigo, 1996). Areas defined by frontiers cannot be observed without taking into account the practices and reality of the actors involved. And yet the journeys taken by migrant are becoming less linear (from the point of departure and the point of arrival) and more complex in terms of time required, as they are forced to retrace their steps, wander in search of alternatives, or settle in transit zones, all the while multiplying and renewing points of departure, transit and arrival.

Finally, this collection of papers highlights the transformation of spatial and temporal perceptions and practices by migrant populations. The introduction of increasingly complex and many-layered processes of control has had a fundamental impact on their experience. The way that they perceive and understand territories varies according to the status of each person, their rights and their ability to travel, whether in the urban area where they live or at international frontier crossings. Further, these differences highlight the emergence of new categories of mobile and immobile, legal and illegal migrants (Bauman, 1999). The establishment of different and specific forms of restriction and mobility contributes to growing differences in the way they experience territorial spaces, because of their personal bio-social profile. When their mobility is restricted or blocked, actors will sometimes react by developing various ruses, skills, tactics (de Certeau *et al.*, 2002) and strategies of adaptation, avoidance, resistance or combat: adopting new itineraries, taking different routes, becoming invisible in order to avoid changes in control points. They are forced to be inventive and innovative in order to overcome geographical obstacles and navigate their way through the various types of status defined by legislation or regulations, and this in turn leads to the presence of new intermediaries at the frontiers (smugglers, mafia, recruitment agencies, producers of false papers, etc. (Andreas, 2001)). Thus new business opportunities emerge for trafficking using informal actors and activities that seek to circumvent sophisticated systems for controlling the passage of migrants and to turn them to their advantage through retroactive dynamics (Shamir, 2005). In their daily life, migrants need to develop new strategies of mobility in order to overcome fear, isolation, and invisibility in response to the location and disposition of control points. They must deal with shifting territories, find new groups that will allow them to integrate the society in which they land, or to discover new opportunities for moving on by developing or making use of renewed transnational links.

Increased transnational circulation of migrants, experienced by those who are sometimes unable to find their place to settle anywhere, challenges current theories about social integration. It questions the too long-dominant paradigm that immigrants can be assimilated and will identify with the diaspora, and invites empirical research into how transnational social fields are constituted by taking advantage of the very abundant literature produced by Anglo-Saxon research since the 1990s (e.g. Basch *et al.*, 1994; Glick Schiller and Wimmer, 2002). The process of memorial reactivation within migrant groups is based on the dynamics of circulation between areas of departure and of settlement and on the process of reshaped identities that are inherent with uncertainty over personal origins.

With this collection of articles, we find ourselves at the interface where we can observe the transformation of frontiers and controls of passage, on the one hand, and the nature of reactions, transgressions, re-appropriations, and innovations used by migrants, societies and the economic opportunities that they generate, on the other.

The originality of the volume is threefold. First, it makes it possible to sidestep the opposition of, on one side, the study of institutional systems and political processes involved in the construction of frontiers and territories to, on the other, the study of the daily practices of ordinary people who must travel differently by showing how the two interact. It is these interactions, whether they represent dialogue or conflict, between 'bottom-up' and 'top-down' dynamics where reconfigurations of territories and methods and practices for their management occur. Second, the dynamism of these processes, as a move towards setting them within the context of the last two centuries, also permits a discussion of the theoretical novelty of processes thought to be long term structures. Finally, and the third originality of our approach, the Mediterranean region in its broadest sense, to which this volume is mainly dedicated, allows us to adopt a perspective that, if not strictly speaking comparative, is at least a strong dialogue in that it offers an opportunity to demonstrate how these processes, as identified as shared forms of comprehension, operate at the local level. Through the studies presented here describing historically and locally different dynamics, the reader will be able to distinguish aspects that are specific to the various local, national and regional contexts identified from those that are specific to the Mediterranean region or, more widely, to manifestations of contemporary globalization. The question is, therefore, how to put into perspective what is 'new' and what is 'global' in these transformations (Beaujard *et al.*, 2009), in order to re-establish the value given to the specificities of their local manifestations, as part of their historical definition, and to avoid the pitfalls of making generalizations without ignoring the theoretical opportunities. Our observation of the daily readjustments by ordinary people demands that we re-evaluate any approach based on 'top-down' diversity as part of a process of globalization that often becomes too engulfing when seen 'from above'.

In order to respond to these aspirations, this volume has been compiled on the basis of a multidisciplinary reflection bringing together historians, geographers, sociologists and anthropologists who have participated in a transversal research programme within the Maison méditerranéennne des sciences de l'homme (MMSH), Aix-en-Provence (France) in 2009-2012. This is thus one of the group's publications.[2] The

[2] The transversal programme MIMED (Places and territories of migrations in the Mediterranean, 19th to 20th centuries) brought together researchers from four MMSH

fifteen contributions presented in this collection have been grouped under four themes which confront, without seeking to be exhaustive, by taking different references in terms of perspectives and temporalities, a variety of local contexts in countries in the Northern (France, Greece, Italy, Spain) and Southern (Morocco, Tunisia) Mediterranean as well as in its Eastern region (Israeli-Palestinian region, Lebanon) and, for the purposes of comparison, the United States.

The first section, *Mobility and Systems of Control*, analyzes changes in institutional practices in controlling frontiers. Moving from a regime of free circulation to one of permanent controls, during the period (1880-1930) of the development of nation-states, these practices have given rise, in Europe since the mid-1990s, to new forms of externalized and negotiated political controls in the Euro-Mediterranean region. These recent changes included systems that became more and more elaborate and which are analyzed in the second section: *Mobility and the Border Economy*. Based on field studies in small local communities, this group of articles seeks to explain how mobile frontier populations bypass and re-appropriate the inherent weaknesses within frontier control systems and how these have an influence on the way these same systems are then readjusted. It also puts into perspective the wide range of institutional and non-institutional, formal and informal actors who are involved in the current operation of control systems at the frontiers of Europe and the Mediterranean. Taking a lateral approach to the urban areas where 'illegal' migrants tend to settle, the third section, *Beyond State borders, migrant's mobility in everyday life*, shows how frontiers have become part of individual migrants' corporal existence and defined by their status (irregular, asylum seekers, clandestine minors, etc.). The practices, commentaries and representations of vulnerable migrants lie at the heart of these contributions which analyze, with their relations with the 'other', the tactics and strategies used to overcome the many legal, symbolic and moral frontiers they face in everyday life and the ways in which they contribute to the reconstruction of identity and

---

laboratories: Institut de recherches et d'études sur le monde arabe et musulman (IREMAM), Institut d'ethnologie méditerranéenne, européenne et comparative (IDEMEC), Laboratoire méditerranéen de sociologie (LAMES) and Temps, espaces, langages, Europe méridionale-méditerranéenne (TELEMMe). Logistical and financial support was provided by MMSH, the 4 laboratories, and 3 partner laboratories (MIGRINTER, ART-DEV, and LEST), several of whose researchers being closely associated with the project, and by the University of Provence and the Provence-Alpes-Côte d'Azur Regional Council. The coordinators also wish to thank the Centre de Recherches Français de Jérusalem (CRFJ) for its support and facilities offered for the international seminar (June 2011) on the theme "Territorial and Border Configurations to the test of Mobility and Migration, 19-21st centuries" during which the idea for this volume was launched.

society. And finally, the role of memorial events in the construction of identity, appropriation and territorial dynamics of countries of origin and of destination is the theme for the last section, *Memory and Migration, Memory of Migrations*, in which different spatial and temporal contexts are examined through the prism of a polyphony of identities invoked by memories of the past.

References:

About I. and D. Vincent 2010. *Histoire de l'identification des personnes*. Paris: La Découverte.

Andreas P. 2001. *Border Games: Policing the US-Mexico Divide*. Ithaca: Cornell University Press.

Audebert C. and R. Nelly 2009. "L'externalisation des frontières des 'Nordsanbsp' dans les eaux des 'Suds'". *Cultures & Conflits*, 73: 35-51.

Baduel P.-R. (ed.) 2009. *Construire un monde? Mondialisation, pluralisme et universalisme*. Paris: Karthala.

Basch L., N. Glick Schiller and C. Szanton Blanc 1994. *Nations Unbound: Transnational Projects, Postcolonial Predicaments, and Deterritorialized Nation-States*. Langhorne: Gordon and Breach.

Bauman Z. 1999. *Le coût humain de la mondialisation*. Paris: Hachette.

Beaujard P., L. Berger, and P. Norel 2009. *Histoire globale, mondialisations et capitalisme*. Paris, La Découverte, Recherches.

Beck U. 2003. *La société du risque. Sur la voie d'une autre modernité*. Paris: Flammarion-Champs.

Bel Air (de) F. (ed.) 2006. *Migration et politique au Moyen-Orient*. Beyrouth: IFPO.

Bigo D. 1996. *Polices en réseaux: l'expérience européenne*. Paris: Presses de Sciences Po.

Bigo D., R. Bocco Riccardo, and J.-L. Piermay 2009. "Logiques de marquage: murs et disputes frontalières". *Cultures & Conflits* [online 28.12.2009], 73 | spring 2009, [Accessed 10.10.2012] URL: http://conflits.revues.org/17484.

Boucheron P. (ed.) 2009. *Histoire du monde au XVe siècle*. Paris: Fayard.

Braudel F. 1993a. *Civilisation matérielle, économie et capitalisme XV-XVIIIe siècle*. Paris: Livre de Poche.

Braudel F. 1993b. *La Méditerranéenne et le monde méditerranéen à l'époque de Philippe II*, Paris: Livre de Poche.

Bredeloup S. and O. Pliez (eds.) 2005. "Migrations entre les deux rives du Sahara". *Autrepart*, 36.

Brown W. 2009. *Murs: les murs de séparation et le déclin de la souveraineté étatique*. Paris: Les Prairies ordinaires.

Certeau (de) M., L. Giard and P. Mayol 2002 [1980]. *L'invention du quotidien, 1/ Arts de faire*. Paris: Gallimard/Folio Essais.

Cuttitta P. 2006. "Points and lines. A topography of borders in the global space". *Ephemera*, 1: 27-33.

Fabre T. and P. Cassia (eds.) 2008. *Between Europe and the Mediterranean. The Challenges and the Fears.* London: Palgrave Macmillan.

Foucher M. 1986. *L'invention des frontières.* Paris: FEDN.

Glick Schiller N. and A. Wimmer 2002. "Methodological nationalism and beyond: nation-state building, migration and the social sciences". *Global Networks*, 2(4): 301-334.

Grataloup C. 2006. *Géohistoire de la mondialisation, le temps long du monde.* Paris: Colin.

Latte Abdallah S. and C. Parizot 2011. "A l'ombre du Mur: comprendre le régime d'occupation israélien", in S. Latte Abdallah and C. Parizot (eds.), *À l'ombre du Mur: Israéliens et Palestiniens entre séparation et occupation.* Arles: Actes Sud: 15-48.

Lévy J. 1999. *Le Tournant géographique. Penser l'espace pour lire le monde.* Paris: Belin.

Noiriel G. 1992. *Immigration et identité nationale en France (XIXe-XXe siècle).* Paris: Hachette.

Parizot C. 2009. "Temporalities and perceptions of the separation between Israelis and Palestinians". *Bulletin du Centre de recherche français à Jérusalem* [Online 10.3.2010], 20 | 2009 [Accessed 10.10.2012 URL: http://bcrfj.revues.org/6319.

Sassen S. 1991. *The Global City.* Princeton University Press.

Schnapper D. 2001. "De l'État-nation au monde transnational. Du sens et de l'utilité du concept de diaspora". *Revue européenne des migrations internationales*, 17(2).

Shamir R. 2005. "Without Borders? Notes on Globalization as a Mobility Regime". *Sociological Theory*, 23(2): 197-217.

Simon G. 2008. *La planète migratoire dans la mondialisation.* Paris, Collin U.

Urry J. 2000. *Sociology beyond Societies: Mobilities for the Twenty-First Century.* New York: Routledge.

Wallerstein I. 2006. *Comprendre le monde, Introduction à l'analyse des systèmes mondes.* Paris, La Découverte, Grands repères.

Wihtol de Wenden C. 2009. *Atlas mondial des migrations. Réguler ou réprimer... gouverner.* Paris: Autrement.

Wolff S. 2008. "Border management in the Mediterranean: internal, external and ethical challenges". *Cambridge Review of International Affairs*, 21(2): 253-271.

PART I

MOBILITY AND SYSTEMS OF CONTROL

Introduction

Sylvie MAZZELLA, Céline REGNARD

Frontiers are not natural barriers, but historical constructions, institutional and political issues that give birth to a specified space (Nordman, 1999). The four chapters in this first section examine the internal and external mechanisms for the political construction of border controls, from the first appearance of the nation-state and of the concept of national identity, in the late 19th century and early 20th century in Europe until now, in the early 21st century, when European identity is being redefined and controls redeployed in the Mediterranean.

Inevitably, the history of changing frontiers includes that of migrants, who are now required to produce identity papers and other documents that will justify the legitimacy of their passage. The installation of frontiers, and their later evolution, not only draws lines around territories but also divides legal travellers from illegal migrants.

The period covered here has been marked by two important changes: the installation of intra- and extra-European frontiers and of the attendant checkpoints; and growth in migratory flows, first of Europeans and later of other nationalities (Withol de Wenden, 2005). These four articles seek therefore to establish connections linking legislation that defines national and territorial affiliations, tighter frontier controls, and the assignation of specific status to migrants that definitively marks them as legal or illegal. Human mobility, a phenomenon much feared by nation-states in the past, has always represented a challenge, but particularly when political frontiers are transformed into barriers separating areas with different economic realities.

In Europe, during the period between the 1880s and 1930, the policy of free circulation was transformed into a regime of permanent controls of identity. From a historical perspective, we can see the emergence of a need to check personal identities and the development of appropriate technology in order to ensure more rigorous controls at border crossings. With these controls, the frontier 'exists' as more than a mere line, barrier or even '*limes*'. Thus, since the 19th century, nationality is now an essential criterion for governments seeking to define social identity. For the historian Gérard Noiriel (2007), this process towards 'nationalization' is indicative of an 'identity revolution'. The real presence of a frontier becomes part of the population's daily life, as demonstrated by Ilsen About, and is associated with the development of a national identity that confirms the country's unity. The author specifically analyzes the impact of the Circular of 1899 in France and the role played by police officers in establishing a national frontier that is not merely a geographical limit, but also a symbolic place that awakens self-awareness. The situation on the Franco-Italian border, as discussed in this article, offers a good illustration of this phenomenon, as it allows us to observe one of the major waves of Mediterranean migration: Italians fleeing poverty and later Fascism in their country for a new future in America and France.

Since the Second World War, two phenomena have contributed to the debate on migration and frontier controls: the very high numbers of migrants travelling around the Mediterranean in response to strong economic development, notably in France; and, the process of decolonization which changed the nature and quantity of migration flows within and towards Europe (Liauzu, 1996), the huge wave of migrants now composed of Portuguese and Spanish, but most of all North Africans. These phenomena show us the importance of discussing frontier controls as a fundamental feature of Europe's operational structures and as a significant political marker.

Over the last thirty years or so, the need for global controls – which the European Union has sought to harmonize – has become a major issue. Border controls are being placed at the EU's outer edges. Implementation of the Schengen Convention (1995) led to the signature of bilateral agreements between EU Member States and countries in the Mediterranean and outside Europe (de Tapia, 2008). At the same time, the introduction of sophisticated technological innovations (biometric passports, walls, drones, etc.) and new resources, such as the Frontex agency, have facilitated the management of cooperative operations at the EU Member States' external borders. The EU's new frontier regime, which reinforces complex political cooperation, nevertheless sustains asymmetric power struggles between the North and the South in terms of migration controls in Europe and in the Euro-Mediterranean region. Thus, Southern European countries, located at the interface with extra-European and Mediterranean zones, are now expected to guarantee the impermeability of the EU's frontiers.

These Southern European countries, weakened by the economic crisis and traditionally countries of emigration as Sandro Rinauro's article shows for Italy, are being transformed against their will into countries of destination for new migratory flows and are required to become gatekeepers for the European Union. The arrival of migrants from Sub-Saharan Africa and Central Europe represent a new challenge.

In her article, Laurence Pillant reviews the security arrangements set up at the Greek-Turkish frontier as part of a 'frontier extension area' and the ways that new actors, whether private or public, national or European, governmental or non-governmental are becoming interdependent. The frontier and its controls appear to be diffracted. In response to the multiplication of agencies and the need for different types of controls, there are now numerous places where identities are checked and which are not necessarily situated at the frontier itself. On the one hand, today as in the past, camps where migrants await a decision authorizing their stay or confirming their expulsion represent and symbolize migration control while, on the other, there are more ordinary and banal controls at the boarding gates for ferries. Control techniques have also changed and are no longer limited to a simple identity check: they now contribute an opportunity for analysis by use of personal interviews with migrants. Yet, although the frontier and frontier controls have become more 'widespread' or at least less visible, this should not be interpreted as a weakening of state authority which, while establishing these new procedures, continues to wield its considerable powers behind a screen of reconfigurations and redeployments.

Sylvie Mazzella's article highlights the way that controls of migration flows between European Union and North Africa can now be compared with those established on the periphery of the European Union. On the one hand, North Africa, once a region of emigration like Greece and Italy, is now a region of immigration. This area has become a transit region, because of differences in economic conditions between Africa and Europe and because of the political events that have shaken the African continent. On the other hand, we have seen that the European Union has introduced a twofold process: it has externalized its border controls and required countries of origin to assume responsibility for their nationals. This has led to a transformation in the relations with countries which, during the colonial or post-colonial environment, were often seen as 'dominated' but are now economic and indispensable partners for the European Union in controlling its frontiers.

Although control systems have changed, as discussed in these articles, there are nevertheless several constants. For example, economic and political issues have become more acute since the 19th century, due to industrialization and the movement of political refugees. When these refugees are a source of cheap labour, as discussed by Sandro Rinauro, this form of labour migration is generally well received, even actively encouraged. Yet, generally speaking, taking a tough political stance on the question of national, or European, identity, together with differences in economic development, has generated a certain amount of distrust towards migrants since the 19th century. The 'false' migrant or asylum-seeker is now seen as a threat at a sensitive period when the nation-state is being redefined and the construction of a European community is in crisis. This distrust reached its paroxysm with the introduction of policies that seek to control or expel peoples who have roamed throughout the continent for more than two hundred years (Asséo, 1994 [2006]). In fact, as these four articles show, criminalizing migrants is not a new phenomenon and changes over time. Introduced gradually during the 19th century, administrative procedures that classify and identify migrants are now taking central stage (Rea, 2000; Palidda, 1999).

Political decisions on control procedures, though diverse and spread throughout the Euro-Mediterranean region, seem nevertheless to be less organized or rational than we might expect and inherent weaknesses are emerging and undermining future actions. Does this suggest, nevertheless, that state sovereignty – with regard to migration controls – is in decline? The authors' reply is no: this is not so much a decline, as a redeployment of the administration's methods for exercising power.

References

Asséo H. 2006. *Les Tsiganes, une destinée européenne*. Paris, Gallimard.

Liauzu C. 1996. *Histoire des migrations en Méditerranée occidentale*. Paris: Complexe.

Noiriel G. (ed.) 2007. *L'identification. Genèse d'un travail d'État*. Paris, Belin.

Nordman D. 1999. *Frontières de France, de l'espace au territoire XVIe-XIXe siècles*. Paris: Gallimard.

Palidda S. 1999. "La criminalisation des migrants". *Actes de la recherche en sciences sociales*. 129: 39-49.

Rea A., F. Brion, C. Schaut, and A. Tixhon. 2000. *Mon origine, mon délit. Criminalité et criminalisation de l'immigration*. Bruxelles: De Boeck.

Tapia S. (de) 2008. *Système migratoire euro-méditerranéen*. Strasbourg: édition du Conseil de l'Europe.

Withol de Wenden C. 2005. *Atlas des migrations dans le monde: réfugiés ou migrants volontaires*. Paris: Autrement.

Illegal Migration of Italians from the Liberal Age to Freedom of Movement Within the European Community

Geographic Directions, Public Attitudes and the Strategies of Migrants

Sandro RINAURO

Illegal expatriation in liberal Italy

Illegal migration was widespread in Italy during the classical era, during the Middle Ages and in pre-Unification Italy, but the scale of the phenomenon and the awareness that migrants and authorities had of it varied over the ages. This was because it was only when public power was strong and able to control the borders that laws on cross-border circulation of subjects and foreigners could be emanated and attempts made to control it (Borruso, 2001; Sanfilippo, 2011; Rinauro, 2009).

Even so, illegal emigration of Italians reached considerable proportions only in the second half of the 19th century. This was because the total exodus came on a massive scale (surpassing 100,000 expatriations per year as from the 1870s) and also because the building up of nation states in the West had led to a consolidation of the legal distinction between citizens and foreigners. Furthermore, the material instruments for ascertaining this difference had become readily available in the form of passports, identity cards with anthropometric data and, later, with photographs. From this time on, as never before, "papers" (expatriation documents) became the real obsession of migrants (Costa, 1999-2001; Torpey, 2000; Green, Weil, 2007; Franzina, 2005). Access to and residence in immigrant-receiving countries was no longer granted on an individual basis, depending on the person's reputation, but rather on the basis of an impersonal "paper identity" in the form of identity and expatriation documents.

In the decades of the "Great Migrations" around the turn of the 20th century, there was however a contradiction between the growing limitations on the free circulation of foreigners imposed by nation states and the first signs of economic globalisation that was leading to enormous migratory movements of workers during the period. In an attempt to encourage worldwide economic integration, international migration laws were actually becoming increasingly liberal, thus making illegal immigration to some extent superfluous. Furthermore, while the conditions and verification required for entering foreign countries became stricter, some of the main destinations for migrants were also beginning to encourage the naturalisation of first-generation foreigners. This was the case in countries that needed to expand their populations, such as the United States of American and the larger Latin American republics, but France too began to promote naturalisation, even though only of foreigners born in the country, with its citizenship laws of 1851 and 1889 (Weil, 2005).

Ultimately, the phenomenon of illegal immigration during the time of the "Great Migrations" at the turn of the 20th century was provoked by the consolidation of the nation state. However, owing to liberal immigration policies and facilitated naturalisation, the status as a clandestine resident became a transitional phase during the life of the migrant, for it concerned only the moment of the border crossing and the first years of residence in the receiving country. In the case of Italian migration, this would appear to be confirmed by the statistics, for in absolute terms there were considerable numbers of illegal immigrants (and the figures are certainly underestimates), and yet in percentage terms they were very few compared with the number of legal migrants.

The first statistics on Italian migration to be compiled in a systematic manner, by Leone Carpi in the late 1860s and early 1870s, give the figure of illegal immigrants in 1870 as 8,643, compared with 101,815 legal immigrants. The number of illegal immigrants was thus a bit less than 8% of the total number of migrants. In 1871, 11,068 illegal immigrants were apprehended out of a total of 122,479 (9%), and in 1873 the figure was 11,833 out of a total of 147,003 (8%) (Carpi, 1874). The author himself stated that these were uncertain, incomplete figures (Rinauro, 2010). In any case, Carpi considered legal migrants to be those who expatriated with a passport and illegal migrants those who were without, or those who immigrated legally for European destinations but then who crossed the Atlantic without a passport for their final destination. The reasons for illegal expatriation mentioned by Carpi were primarily failure to report for military service – of the 642 illegal emigrants apprehended in 1873 in Genoa as they were about to leave, 522 had failed to report for military service and 30 were wanted for other offences. Secondly, there was the lack of funds to pay for the ocean crossing and, lastly, the lack of a work contract abroad, without which the Italian authorities would not issue a passport. In those years, the legal restrictions that led to illegal emigration were thus not so much due to immigration policies in the country of destination, which as we have seen were liberal, so much as to hostility towards the exodus in Italian law and, more in general, among the ruling classes in Italy. The ban on expatriation for those who could not demonstrate a job contract abroad or, alternatively, sufficient means of sustenance, had been introduced by the circular issued in 1868 by the President of the Cabinet Council, Menabrea, and confirmed by the circular issued by his successor, Giovanni Lanza. Dated 18 January 1873, the latter also required, as a condition for the release of a passport, an undertaking to reimburse the Italian state in the case of repatriation at the expense of the consulates (Ostuni, 2001; Manzotti, 1962).

Also Italian family law at the time tended to encourage illegal immigration by hindering the expatriation of adults suspected of avoiding their family responsibilities, of minors without parental consent, and of women who, emigrating alone or without the legal consent of their father or husband, were suspected of entering the international prostitution market. Furthermore, sporadic operations by the Italian authorities to prohibit expatriation to countries or regions where the treatment of immigrants was particularly inhumane led by reaction to illegal immigration, as was the case in 1911 towards the La Plata countries. At any rate, the immigrant-receiving countries that would later become increasingly habitual for Italian illegal migrants were becoming established even then. France and Switzerland topped the list, with many other countries in Europe, since they were on Italian borders that were difficult to control, but also the Americas, despite the apparent ease of verification at embarkation and disembarkation ports (Frascura, 1904; Sori, 1979; Cioli, 1989; Porcella, 1998). At the same time, criminal migrant-smuggling organisations and document forging workshops were already appearing in these decades. At the turn of the century, the Sicilian Mafia was forging passports and smuggling illegal migrants to New York via Mazara del Vallo, Tunisia and Marseille (Lupo, 1996, 153-154). In these years there was also the "white slave trade" and the notorious trade in Italian children for French glassworks, where sickness and mortality rates were extremely high. These were already defying the law and heralding far greater criminal speculations at the expense of illegal migrants (Paulicci de Calboli, 1902; Protasi, 1999; Bianchi, Andrea, 2000; Stella, Franzina, 2001).

After the emigration law of 1888, the phenomenon of illegal migration grew even further, not just in relation to the constant increase in expatriations but also as a result of the law itself. With the aim of protecting emigrants from fraudulent recruiters, countering their migration propaganda, and assisting Italian steamship companies to the detriment of foreign ones, this law made a licence compulsory for emigration agents (independent recruiters who were paid by any shipping company, whether Italian or foreign) and for carriers (agents employed by a single shipping company). From that moment on, and even more after the emigration law of 1901, which abolished the figure of the independent emigration agent and allowed only that of the carrier, Italian law and statistics considered as illegal not only those who expatriated without a passport, but also those who had been recruited by unlicensed agents and carriers, or who embarked at foreign ports or with foreign shipping companies that were not authorised by Italy for the transport of migrants (Martellini, 2001). The reason why emigrants turned to illegal brokers was because they could obtain tickets at much lower prices then they could from Italian shipping companies, but also because illegal recruitment agencies would disembark them also in countries that were temporarily closed to Italian migrants (as a result of bans by Italy following episodes of excessive exploitation of immigrants or during epidemics), or where quotas were in force to limit entry.[1] As a result, in order to elude the Italian authorities, who did not allow them to be recruited by illegal brokers, many expatriated legally to other European countries, from which they embarked illegally from these foreign ports on foreign steamships. The most common European destinations for illegal migrants heading for the Americas were thus the ports of Le Havre, Bordeaux, Bremen, Hamburg and Trieste, where their illegal passage was encouraged by foreign shipping companies, who even sent their own unlicensed recruiting agents to Italy. The most active illegal recruitment agencies worked in the main ports of Italy, but they mainly worked out of Switzerland and France.

It was partly due to their smuggling activities that the figures for illegal expatriation from Italy from the end of the 1870s grew so fast, involving at least 20,000-30,000 individuals each year (Sori, 1979, 320-324). However, in the first fifteen years of the 20th century, annual expatriations rose to as high as 600,000, so statistically confirmed illegal migration had actually not increased in percentage terms, accounting for about 5% of total expatriations. The scale of illegal seasonal migration in the north and over the Alps was however incalculable, for these migrants had little care for passports and frontiers in their constant coming and going. As Augusta Molinari wrote about the migrants who went from western Liguria to the south-east of France in the late nineteenth and early 20th century, "'Crossing the border' to do seasonal jobs in France was a centuries-old tradition among the people in the area. Coming and going without any certification from the authorities was customary practice" (Molinari, 1995, 142).

---

1 Commissariato generale per l'emigrazione, *L'emigrazione italiana dal 1919 al 1923*, CGE, Rome 1926, pp. 347-69.

Recruitment by unauthorised agents and carriers often involved the migrants being swindled and, especially when they had crossed the ocean, they would be either exploited because of their status as "undocumented migrants" or they would fail to find work, and they could also be expelled by the authorities, who feared that they had no passport because they were criminals. In order to avert these risks, Law No. 1075 of 2 August 1913 on the protection of emigrants increased the powers of control and investigation by the general emigration commission (Commissariato Generale per l'Emigrazione – CGE) against illegal recruitment. This greater vigilance was also prompted by the introduction of the US "quota laws" of 1921 and 1924, which radically reduced access to what had been until then the main destination of Italian migrants. This led to an even greater wave of illegal migration and an enormous rise in the price demanded by recruiters. French, Swiss and Belgian recruitment agencies intensified the enrolment of illegal Italian migrants, promising them easy entry into the USA from Mexico, Canada and Cuba. This led to a rise in the number of migrants intercepted by the CGE and by Italian police on the Alpine borders as they made their way to the foreign agencies in order to embark illegally from ports in France and Belgium. Networks of Italian recruiters working for foreign agencies were thus discovered but, due to a lack of personnel and means, the CGE had to admit that a great deal of illegal migration managed to elude their control. The numbers of Italians who left from foreign ports illegally (without a passport) were thus incomplete but nevertheless conspicuous: 26,801 Italians, at least 8,000 of whom were illegal emigrants, reached the United States from the port of Le Havre in 1912, 6,187 in 1913, and at least 10,565 Italians entered Brazil illegally in 1910 and 1911. In the years around 1907 at least 20,000 Italians crossed the Italian-Swiss border on their way to Swiss and French recruitment agencies who would then smuggle them across the ocean.[2]

Illegal migrants naturally also boarded ships at Italian ports, where fake forwarding companies helped them on board by disguising them as sailors or porters with the bribed connivance of the crew, and sometimes even of the captain, as well as forging passports for them. Stowaways were often discovered and reported to the CGE agent on board, but many of them managed to land on the other side of the ocean (Franzese, 1996; Masi, 2003). Partly as a result of attempts to go abroad by deserters and absentee conscripts during the First World War, in the 1910s the CGE intensified its preventive and law-enforcement activities. Specialised CGE staff monitored the main Alpine passes and the four Italian emigration ports. They were assisted in this by the police force and the

2 *Idem.*

Carabinieri, and in 1911 a land-border office (Ufficio per i Confini di Terra) was set up in Milan, together with local offices in Bologna, Udine and Turin. These had the task of checking expatriation documents of migrants at Italy's land borders and to inform them about the foreign labour market and monitor the activities of Italian and foreign emigration agents. At the same time, the CGE requested consuls in areas near European ports for ocean crossings to provide data concerning illegal boarding by all nationalities and for the identity of the emigration agents involved, and it introduced an internal index of illegal agents and recruiters. These multiple inspection activities enabled the commission to present 4,431 complaints to court against 5,808 people, most of whom were illegal recruiters and swindlers of migrants. Sentences, which were fairly light, were however handed down to fewer than half of the accused. In the commission's view, this was due to the excessive leniency of magistrates who were unaware of the dramatic situation faced by stowaways and swindled migrants.[3]

Illegal migration between the wars

The illegal exodus was huge even during years of liberal migration policies, so it could hardly fail to explode when, from the 1920s, the first blanket restrictions came into force with the introduction of ethnic "quotas" for access, which were adopted by the United States as well as by Canada and Brazil (Divine, 1957; Higham, 1965; Martellone, 1980; Hutchinson, 1981; Tichenor, 2003). It was mainly the Great Depression that drastically reduced expatriation, but in the late 1920s also the Fascist regime in Italy adopted an extremely restrictive emigration policy, in line with its demographic and military policies (Cannistraro, Rosoli, 1979; Labanca, 2000; Luconi, 2000; Bertonha, 2001; Franzina, Sanfilippo, 2003). In reaction to this, illegal emigration shot up, especially in relation to the small total number of emigrants. The political exile of many anti-fascists also had an unprecedented impact on the flow of illegal emigrants. Their number reached 60,000 individuals (Audenino, Tirabassi, 2008; Tombaccini-Villefranque, 1999). These emigrants were often illegal in the eyes of Fascist law, which attempted to prohibit their expatriation with the introduction of Law No. 1278 of 24 July 1930, but they were equally so for the destination countries which, in years of widespread reaction, did not grant all of them political asylum. When the racial laws were introduced, the exodus of anti-Fascists was also joined by that of Italian Jews and of Jews passing through Italy (Veziano, 2002; Istituto storico della Resistenza e della società contemporanea in provincia di Cuneo, 2007).

---

3 *Ibid.*, p. 360-368.

These restrictions on migrants in the inter-war years also sparked off the illegal emigration of workers, as the extremely high number of illegal Italian immigrants in France in particular clearly shows: between 1920 and 1933, next to an average annual inflow of about 10,000-15,000 legal Italian immigrants, another 20,000-25,000 illegal and irregular Italians entered the country each year.[4] Already common in the 1880s, the number of illegal Italian immigrants in France grew after the First World War with the introduction of more restrictive recruitment policies by the French and, in the opinion of the police in both countries, it became uncontrollable (Maltone, 1988, 103-118; Tombaccini-Villefranque, 1999, 80). The Italians especially entered as false tourists and pilgrims, and using other stratagems, and then when they found work they would try to get their papers in order (Demangeon, Mauco, 1939).

Under pressure from local employers, the French authorities were so tolerant that no fewer than a third of foreign workers who arrived in France in the 1920s were illegal immigrants. Of the 221,619 foreign workers who arrived in 1930, for example, a full 60,000 had entered the country illegally, and they were mainly Italians, Spanish and Belgians (Henneresse, 1978). The 1929 Depression, however, also discouraged illegal entry (Maltone, 1988, 116-117).

Illegal emigration during the years of the reconstruction

Immediately after the Second World War, the illegal flow of political emigrants – war criminals, Fascist party officials, particularly ruthless republican Fascists and public and private managers who had collaborated in the war economy of the regime and of the Nazi occupiers – was important in political terms but involved just a few hundred people. Their destinations were mainly in Latin America and Switzerland, from which many returned to benefit from amnesties and frequent acquittals (Bertagna, 2006; Van Dongen, 2008). Under the Nazi occupation of Italy, some thousands of Italian Jews illegally fled abroad in the autumn of 1943, and those Jews – both Italian and foreign – who in the immediate aftermath of the war sailed from Italy for Israel before its constitution as an independent nation, were also illegal emigrants (Toscano, 1990; Sereni, 1994).

It was in the post-war period that illegal labour emigration really took off, more than previously, not just because of the disproportion between the countless Italian candidates for work abroad and the limited need for manpower in the receiving countries, but especially because of

4 Centre des Archives Contemporaines, Fontainebleau (CAC), Pos. 19900544, Art. 3: programme d'immigration.

the red tape and the restrictive, bureaucratic nature of immigration policies in these countries (Rinauro, 2005).

After the mass conversions to Fascism and Communism in the 1930s, in the post-war period the Western democracies found "full employment" and anti-cyclical economic policies to be the fundamental tools for maintaining international peace and democratic consent at home. This led to the international adoption of free circulation, on which Italy had insisted, being replaced by resort to so-called bilateral migration treaties with which immigrant-receiving countries imposed the quantity, quality, employment sector and short duration of immigration. These countries also established the sole place of work and residence and even the size of the remittances that immigrants could send back to their families in Italy. Once they were abroad, foreigners who changed the district of residence assigned to them or their profession or even only their employer, or who stayed beyond the terms of their contract, became illegal immigrants and were thus immediately repatriated, at times after spending time in prison. The process of setting up such a precise system for managing foreign labour proved to be highly complex and long, both for the migrants recruited directly by the states – in the form of "assisted migration" – and for those recruited individually by foreign entrepreneurs. The procedure took between 2 and 6 months, which meant that sometimes regular migrants would reach their destination when their work was no longer needed.[5] Ignorance of the bureaucratic procedure for recruitment, its excessive length and the constraints of place and profession imposed abroad encouraged large numbers of Italians to avoid the public recruitment process and to expatriate illegally, or to change their profession and place of residence from those assigned to them, which meant they were abroad unlawfully.[6]

What is more, even when demand for foreign workers was high, the host countries often hindered family reunions, both to prevent workers from becoming rooted in the country and to maintain a high percentage of economically active people in the population. Very often the result was that reunions took place illegally. Coupled with this was the interest of foreign business owners, and sometimes also of their governments, to receive illegal immigrants who were thus docile, cheap and precarious and, in any case instantly adaptable to the economic situation.

---

5 On these mechanisms, see the text of the main migration agreements, published at the time in the *Atti della Camera dei Deputati*, the annuals of the *Bollettino quindicinale dell'emigrazione* of the Società Umanitaria of Milan (from 1947), and of *Italiani nel mondo*, from 1945. CAC, Pos. 19810201, Art. 7: Direction de la Réglementation du ministère de l'Intérieur au secrétaire d'État au Travail, Paris, 24 April 1957.

6 "L'emigrazione clandestina in Francia", *Bollettino quindicinale dell'emigrazione*, II (1948), Nos. 16-17, p. 331.

The flow of illegal migrants across the oceans started up again and also led to bereavements, but more frequent still was illegal migration made possible by Italian and foreign swindlers who procured false work contracts, especially for land settlements, building and transport. Due to the lack of verification about the reliability and legality of the recruiters, passport offices issued papers to recruits who then found themselves without a job or even the means to return home once they reached their destination. In countless cases, there were also migrants who, without a work contract or a "letter of call" from relatives, went overseas with just a tourist passport and then stayed on illegally. The Società Umanitaria estimates that these were indeed a "majority" of those Italians who migrated to Latin America between 1946 and 1956.[7]

In any case, it was mainly to European countries that illegal migrants headed, as they were geographically closer, informal communications were easier, their immigration policies were restrictive and they were called by entrepreneurs, but especially because the public authorities were very tolerant. Here the main destinations were France, Switzerland, Belgium and Luxembourg and, from the late 1950s, also West Germany. In the case of Belgium, there were many who had entered France legally or illegally and who then went to Belgium without permits in search of a job. Together with the illegal recruitment by swindlers and false recruitment agencies, it was more frequently Belgian companies that recruited in Italy directly, eluding checks by the countries concerned in order to find cheaper, more flexible manpower. However, it was often the immigrants themselves who put themselves into a state of illegality by abandoning their most common job as underground workers in coal mines, in order to avoid the frequent mortalities and silicosis. Those who left the mines were imprisoned and then repatriated.[8]

During periods of greatest demand, however, it was especially the collieries, or *charbonnages*, themselves that sent their own recruiters to Italy or to mining areas in France to hire illegal Italian migrants. The most disturbing episode came just after the tragedy of Marcinelle where, in August 1956, 136 Italians died. After the accident, the Italian Government suspended recruitment of miners for Europe, while waiting for

---

7 "Emigrazione clandestina", *Bollettino quindicinale dell'emigrazione*, X (1956), No. 24, p. 363.

8 Archivio centrale dello stato, Ministero del Lavoro e della Previdenza sociale, Direzione generale collocamento della Manodopera, Divisione accordi di emigrazione verso paesi comunitari, busta 364, fascicolo 1 "Emigrazione italiana in Belgio – Informazioni e notizie": R.N. Gentile a Mlps, Foggia, 25 ottobre 1946 and busta 386, "Operai italiani clandestini in Belgio": Ambasciata d'Italia a Bruxelles a Ministero Affari esteri, Direzione generale emigrazione Ministero del Lavoro e della Previdenza sociale, Servizio Migrazioni, all'ambasciata d'Italia a Parigi, ai consolati di Liegi e Charleroi e alla Cancelleria consolare di Bruxelles, 21 maggio 1948.

the ECSC safety code for mine work, but the collieries got round the prudent measures of the Italian authorities by hiring illegal workers en masse (Rinauro, 2009, 115-124).

As in the early 20th century, in Switzerland there were still many private recruitment agencies who provided migrants with work contracts in exchange for money. At times the system was still entirely fraudulent, but in others it led to authentic jobs, which however violated contractual principles. However it was mainly Swiss companies that recruited directly in Italy without respecting the recruitment and labour conditions agreed upon with Rome. In 1954, the Ministry of Labour ascertained that irregular recruitment by Swiss companies constituted "the highest contingent of Italian migrants to Switzerland". The other most significant cause for migration and illegal residence was the extremely restrictive immigration policy: to avoid "foreign infiltration" ("*überfremdung*") into the national population, the Confederation favoured seasonal labourers, hindering permanent residence and family reunions. A significant number of illegal immigrants were therefore women and children. And, what is more, large numbers were Italian workers with just tourist passports who were allowed in by the Swiss border guards. When the Confederation decided to block the influx of illegal immigrants and "false tourists" in 1965, it was seen that, together with 1,500 regular Italian workers per day, there were at least 700 illegal workers and "false tourists" per day who were blocked at entry or expelled from within (Rinauro, 2009, 124-135; Ricciardi, 2010).

The mass exodus of Italians to the German Federal Republic started only in the late 1950s, so the illegal arrival of Italian immigrants started late and, especially, ended early. This was due to the introduction in the 1960s of free movement of European workers within the Community, which made it unnecessary to have a work contract before entering (Romero, 1991, 104-108). Even so, it continued for some years and consisted both of illegal immigrants directly from Italy, and of Italians resident in neighbouring European countries. The presence of women, who were mainly sought for domestic work, was considerable (Rinauro, 2009, 112-115).

In any event, it was in France that illegal Italian immigration reached its highest number, and this was due mainly to the more than tolerant attitude of the French authorities. Between 1946 and the end the of the 1950s, over 50% of workers from Italy entered illegally or as "false tourists", and over 90% of the families who joined them in France did so illegally. There were many reasons for such an extensive flow of illegal migrants: during negotiations for the first migration agreements, Paris accepted the illegal immigrants en masse in order to oblige Rome to accept inferior conditions for legal immigrants in the treaties. Until 1948

Rome prohibited recruitment of emigrants by name to prevent France from taking its most qualified workers, who were most useful for the reconstruction. This meant that French businesses resorted to recruiting them illegally. There were also "ethnic engineering" objectives that induced the French authorities to encourage illegal arrivals: in 1947 Algeria was annexed to the metropolitan territory and obtained freedom of immigration into France as well as priority over foreign immigrants in access to work. Since it could not prevent this process legally, Paris decided to discourage the influx of Algerians, considering them "unassimilable", and allowing illegal Italian immigrants to saturate the market. As for the enormous illegal influx of the Italians' families, the law allowed family reunions only if there was suitable accommodation: due to the scarcity and appalling quality of the housing for immigrants, many Italians were obliged to bring in their families in violation of the law. What is more, through its official recruiting service the state-controlled economy in France imposed the profession, employer and sole district of residence, so many Italians preferred to enter illegally in order to choose their own job freely.

Nevertheless, the causes that mostly led the French authorities to encourage illegal immigration were the bureaucratic delays in official recruiting and the terrible living and work conditions for legal immigrants: because of these conditions, regular immigration was far lower than what was required by the migration treaties, so Paris decided to make up for the lack of regular immigrants by taking in illegal ones. In the end, the great willingness of Paris to accept permanent immigrants and their families, and tolerance of illegal immigration became the means with which it competed with other immigrant-receiving countries in getting its hands on foreign workers.

Demand for manpower and the migration of Italians to France reached its height at the start of the European "economic miracle", between 1956 and 1958, and it was paradoxically at that very time that illegal immigration from Italy reached its apex: of the 80,000 Italians who arrived in 1957, almost 36,000 – in other words, 45% – entered illegally or as regular "tourists". This entry through the side door of illegality had an extremely high cost, however: apart from the insecurity and exploitation to which many illegal immigrants were subjected, crossing the Alps on foot cost many Italians their lives, and women and children in particular, and over 1,000 Italians died in the Indochina War. Most of these were illegal immigrants who had been convinced to join the Foreign Legion by the threat of repatriation for illegal immigration (Rinauro, 2005; Gastaut, Hanus, 2001).

References

Audenino, Patrizia, Tirabassi, Maddalena. 2008. *Migrazioni italiane. Storia e storie dall'Ancien régime a oggi*. Milan: Mondadori: 79-83.

Bertonha, João-Fabio. 2001. "Emigrazione e politica estera: la 'diplomazia sovversiva' di Mussolini e la questione degli italiani all'estero, 1922-1945." *Altreitalie*, XIII No. 23: 39-61.

Bertagna, Federica. 2006. *La patria di riserva. L'emigrazione fascista in Argentina*. Rome: Donzelli.

Bianchi, Bruna, Lotto, Andriana. 2000. *Lavoro ed emigrazione minorile dall'Unità alla Grande guerra*. Venezia Ateneo: Veneto.

Bianchi, Bruna. 2002. "Percorsi dell'emigrazione minorile", in *Storia dell'emigrazione italiana*, Vol. II *Arrivi*, ed. by Piero Bevilacqua, Andreina De Clementi, Emilio Franzina. Rome: Donzelli: 360-363.

Borruso, Paolo. "Note sull'emigrazione clandestina italiana (1876-1976)". *Giornale di storia contemporanea*, IV No. 1 (2001): 141-51.

Cannistraro, Philip V. and Rosoli, Gian-Fausto, 1979. "Fascist Emigration Policy in the 1920s. An Interpretative Framework". *International Migration Review*, XIII No. 4: 673-92.

Capuzzi, Lucia. 2006. *La frontiera immaginata. Profilo politico e sociale dell'emigrazione italiana in Argentina nel secondo dopoguerra*. Milan: F. Angeli.

Carpi, Leone. 1874. *Delle colonie e dell'emigrazione d'italiani all'estero*. Milan: Editrice Lombarda.

Cioli, M.G. 1989. "Il passaporto falso. Vagabondi, clandestini, renitenti in alcuni documenti della prefettura di Genova". *La via delle Americhe. L'emigrazione ligure tra evento e racconto*, Genoa: Sagep.

Costa, Pietro. 1999-2001. *Civitas. Storia della cittadinanza in Europa*. Rome-Bari: Laterza.

Demangeon, Albert, Mauco, Georges. 1939. *Documents pour servir à l'étude des étrangers dans l'agriculture française*. Paris: Hermann & Cie: 628.

Divine, Robert A. 1957. *American Immigration Policy, 1924-1952*. New Haven: Yale University Press.

Franzese, Paolo. 1996. "L'emigrazione negli Stati Uniti d'America ai primi del '900 attraverso i documenti della Questura di Napoli, conservati nell'Archivio di Stato", in *Stati Uniti a Napoli: rapporti consolari 1796-1996* ed. by Daniel Spikes, 129-157. Naples: Filema: 129-157.

Franzina, Emilio and Sanfilippo, Matteo (eds.). 2003. *Il fascismo e gli emigrati. La parabola dei Fasci italiani all'estero (1920-1943)*. Rome-Bari: Laterza,.

Franzina, Emilio. 2005. *'Varcare i confini': viaggi e passaggi degli emigranti. Il caso italiano e le teorie transnazionali*, in *Confini. Costruzioni, attraversamenti, rappresentazioni*, ed. by Salvatici Silvia. Rubettino: Soveria Mannelli : 126-127.

Frescura B. 1904. *Dell'emigrazione clandestina italiana diretta ai Porti Esteri e di alcuni mezzi pratici che potrebbero essere adottati per frenarla*. Genoa: Montorfano.

Gastaut, Yvan, Hanus, Philippe. 2011. "Migrants italiens dans les Alpes françaises après 1945. Une mobilité sous surveillance?". *La Pierre et l'Écrit*, XXII No. 22: 183-207.

Green, Nancy, Weil, François (eds.). 2007. *Citizenship and Those Who Leave: The Politics of Emigration and Expatriation*. Urbana IL: University of Illinois Press

Henneresse, Marie-Claude. 1978. *Le patronat et la politique française d'immigration 1945-1975*. Paris: Institut d'Études Politiques.

Higham, John. 1965. *Strangers in the Land. Patterns of American Nativism*. New York: Atheneum.

Hutchinson, Edward Prince. 1981. *A Legislative History of American Immigration Policy 1798-1965*. Philadelphia: University of Pennsylvania Press.

Istituto storico della Resistenza e della società contemporanea in provincia di Cuneo (ed.). 2007. *Ebrei in fuga attraverso le Alpi*, (CD-ROM).

Labanca, Nicola. 2000. "Politica e propaganda: emigrazione e fasci all'estero" in *Fascismo e politica di potenza. Politica estera 1922-1939*, ed. by Enzo Collotti. Scandicci: La Nuova Italia: 137-172.

Luconi, Stefano. 2000. *La "diplomazia parallela". Il regime fascista e la mobilitazione politica degli italo-americani*. Milan: F. Angeli.

Maltone, Carmela. 1988. "L'introduction de la main-d'œuvre italienne en France, entre les deux guerres". *L'immigration italienne en France dans les années 20. Acte du colloque franco-italien, Paris 15-17 octobre 1987*, 103-118. Paris: Cedei.

Manzotti, Fernando. 1962. *La polemica sull'emigrazione nell'Italia unita. Fino alla prima guerra mondiale*. Milan: Dante Alighieri.

Martellini, Amoreno. 2001. "Il commercio dell'emigrazione: intermediari e agenti.", in *Storia dell'emigrazione italiana*, Vol. I, *Partenze*, ed. by Piero Bevilacqua, Andreina De Clementi, Emilio Franzina. Roma: Donzelli: 303-308.

Martellone, Anna-Maria (ed.). 1980. *La questione dell'immigrazione negli Stati Uniti*. Bologna: Il Mulino.

Masi, Giuseppe. 2003. "Tra spirito d'avventura e ricerca dell''agognato peculio': linee di tendenza dell'emigrazione calabrese tra Ottocento e Novecento", in *Emigrazione e storia d'Italia*, ed. by Matteo Sanfilippo. Cosenza: L. Pellegrini: 115-132.

Molinari, Augusta. 1995. "Storia e storie di emigrazione dal Ponente Ligure. Alcuni percorsi di ricerca". *Recherches régionales*, XXXV No. 3: 137-159.

Ostuni, Maria-Rosaria. 2001. "Leggi e politiche di governo nell'Italia liberale e fascista" in *Storia dell'emigrazione italiana*, Vol. I, *Partenze* ed. by Piero Bevilacqua, Andreina De Clementi, Emilio Franzina. Rome: Donzelli: 309-310.

Paulucci de' Calboli, Raniero. 1902. "La tratta delle ragazze italiane." *Nuova antologia*, XXXVII No. 727: 418-438.

Porcella, Marco. 1998. *Con arte e con inganno. L'emigrazione girovaga nell'Appennino ligure-emiliano*. Genoa: Sagep.

Protasi, Maria-Rosa. 1999. "I fanciulli italiani nelle vetrerie francesi: emigrazione e tratta minorile nel circondario di Sora agli inizi del Novecento." *Studi emigrazione*, XXXVI No. 134: 194-242.

Rapone, Leonardo. 2008. "Emigrazione italiana e antifascismo in esilio." *Archivio storico dell'emigrazione italiana*, IV No. 1: 53-67.

Ricciardi, Toni. 2010. "I figli degli stagionali: bambini clandestini.", *Studi emigrazione*, XLVII No. 180: 872-85.

Rinauro, Sandro, 2005. "Politica e geografia dell'emigrazione italiana negli anni della ricostruzione.", in *L'Italia alla metà del XX secolo. Conflitto sociale, Resistenza, costruzione di una democrazia*, ed. by Luigi Ganapini, 247-84. Milan: Guerini: 247-284.

Rinauro, Sandro. 2010. "Le statistiche ufficiali dell'emigrazione italiana tra propaganda politica e inafferrabilità dei flussi." *Quaderni storici*, XLV No. 2: 393-418.

Rinauro, Sandro. 2005. "Percorsi dell'emigrazione italiana negli anni della ricostruzione: morire a Dien Bien Phu da emigrante clandestino.", *Altreitalie*, XVII No. 31: 4-48.

Rinauro, Sandro. 2009. *Il cammino della speranza. L'emigrazione clandestina degli italiani nel secondo dopoguerra.* Turin: Einaudi.

Romero, Federico. 1991. *Emigrazione e integrazione europea 1945-1973.* Rome: Edizioni Lavoro.

Sanfilippo, Matteo. 2011. "La clandestinità è una storia vecchia: note su alcuni aspetti dell'emigrazione irregolare di italiani". *Giornale di storia contemporanea*, XIV No. 2: 227-236.

Sereni, Adda. 1994. *Clandestini del mare. L'emigrazione ebraica in terra d'Israele dal 1945 al 1948.* Milan: Mursia.

Sori, Ercole. 1979. *L'emigrazione italiana dall'Unità alla seconda guerra mondiale.* Bologna: Il Mulino.

Stella, Gian-Antonio and Franzina, Emilio. 2002. "Brutta gente. Il razzismo anti-italiano", in *Storia dell'emigrazione italiana*, Vol. II *Arrivi*, ed. by Piero Bevilacqua, Andreina De Clementi, Emilio Franzina. Rome: Donzelli: 293-295.

Tichenor, Daniel. 2003. *Dividing Lines. The Politics of Immigration Control in America.* Princeton: Princeton University Press.

Tombaccini-Villefranque, Simonetta. 1999. "La frontière bafouée: migrants clandestins et passeurs dans la vallée de la Roya (1920-1940)" *Cahiers de la Méditerranée*, XXXII No. 58: 79-95.

Torpey, John C. 2000. *The invention of the passport. Surveillance, citizenship and State.* Cambridge: Cambridge University Press.

Toscano, Mario. 1990. *La porta di Sion. L'Italia e l'immigrazione clandestina ebraica in Palestina (1945-1948).* Bologna: Il Mulino.

Van Dongen, Luc. 2008. *Un purgatoire très discret. La transition "helvétique" d'ancien nazis, fascistes et collaborateurs après 1945.* Paris: Perrin.

Veziano, Paolo. 2002. *Ombre di confine. L'emigrazione clandestina degli ebrei stranieri dalla Riviera dei Fiori verso la Costa Azzurra (1938-1940).* Pinerolo: Alzani.

Weil, Patrick. 2005. *Qu'est-ce qu'un Français? Histoire de la nationalité française depuis la Révolution*. Paris: Gallimard.

Building Lines between Nations

Border-making and Police Practices at the French-Italian Frontier, 1890-1914

Ilsen ABOUT

Few borders in the Mediterranean have been studied, examined or described as often – and few as bitterly disputed and crossed by marching armies – as the short line between mountain and coast that separates France from Italy. A close reading of this history provides insight into the almost mythical features and specific social structures that surround the *Passage des Alpes* and sheds light on the series of upheavals that this border underwent until the mid-20th century (Fontaine, 2003; Thoral, 2007; Raviola, 2010). The inter-war period, in particular, has attracted considerable attention as it was a very complex period when several trends co-existed: labour migration to France, political exile, diplomatic rivalries and a greater presence of security forces in order to supervise the frontier (Panicacci, 1992; Vogel, 1997; Mourlane, 1999; Baycroft *et al.*, 2001). Above all, this period made it possible to analyze a territory just before this frontier was once again at the heart of a rapid succession of occupations and of conquests during the Second World War (Panicacci, 2010; Sica, 2011). Meanwhile, the last part of the 19th century has been seen as a watershed, during which the newly formed French Republic (1870) and the newly united Kingdom of Italy (1861) jointly, but not without difficulty, devoted a considerable amount of energy to defining the limits of their territory and to strengthening their frontiers (Bouisset, 2002).

Analysis of this process in the post-1880 period has focused on the impact of diplomatic tensions, implications for the international scene, and the importance of long-term economic rivalry during a period of uncertainty and vigilance between the two countries (*Les Alpes Maritimes*, 1992; Ortolani, 2008). In this context, the level of militarization in borderlands increased spectacularly during the 1880s and 1890s. The transformation of the frontiers shared by France and Italy became increasingly visible: fortified military posts, equipped with garrisons of territorial defence forces, were scattered throughout the valleys and their presence constituted an impressive show of arms on both sides of the mountain passes (Jauffret, 1988; Bottin, 1988). This military presence was reinforced by frontier surveillance practices and resources, which were necessary not only for reassuring the local population and for demonstrate the authorities' intentions but also for establishing new restrictive measures for controlling nationals of the neighbouring country.

Resources for border surveillance were put in place gradually in order to create a balance in the activities of various police services already present in the departments. In addition to the presence of special police superintendents and inspectors in these strategic towns, appointed during the Restoration period (1815-1830), a 'border service' focusing on counter-espionage was established during the 1880s under the authority of the Ministries of the Interior and of War. It was however the army that initially played a preponderant role in the service's operations in collaboration with gendarmes and police officers for many of these surveillance activities (Laurent, 2009: 374-407). With the intensification of these activities after 1890, it was necessary to produce reports which would ensure a permanent and broad knowledge of military and espionage activities. These reports were in fact one of the drivers for establishing an information service which gradually gave rise to the emergence of a category of individuals known as 'suspects'. Initiated in 1886, specific records on these individuals, known as the 'Carnet B', constituted a card-index which contained contributions from the police services but was maintained by the army (Becker, 1973; Baker, 1978).

In 1899, when the military authorities was being discredited by the Dreyfus Affair, a decree created a *Contrôle général chargé de la surveillance du territoire* (General Control responsible for Territorial Surveillance) within the Ministry of the Interior that was to bring together all the services involved in counter-intelligence and border surveillance. The objective was to install officially 'a vast surveillance network' covering all departments located at the country's external frontier in order to ensure continuous border controls based on activities of these security forces which were supposed to be coordinated.[1] This circular came into effect quickly in the French departments along the Italian border where their territories were divided into sectors. Each sector was then managed by special superintendents who reported to the Prefect and coordinated the activities not only of 'correspondents' and civil informers but also of public servants, including customs officers, rural policemen, forestry workers and postmen. The list of names was

[1] Decree dated 20.8.1899, "creating the position of General Controller responsible for supervision of the territory" *Bulletin officiel du Ministère de l'Intérieur*, 9, 1899, 153-154.

then established in order to allocate newly hired personnel to take on surveillance tasks.[2]

The objective of this chapter is to analyze precisely the impact of the 1899 circular and study the activities of police forces and their contribution to the construction of a border that was both a geographical barrier and a symbol and, as such, would have a substantial and simultaneous impact on the landscape and the mind. The surveillance network was in fact developed on the basis of a multifarious 'suspect' whose identity was constructed from a multitude of imprecise elements (Bayon, 2004). Here the objective is to define the decidedly vague characteristics of this figure and examine the development of identification procedures in the borderlands, a process which seemed to be inspired by a general atmosphere of suspicion. Behind these control systems, as administrated by agents in order to meet their objectives, emerges a newly-drawn border which appears, by various techniques, as a new physical barrier. Surveillance officers, regardless of rank and service, travelled the length and breadth of these border regions and could challenge everyone they met to produce proof of their identity and the reasons for their presence. Little by little, the border became a real entity in the daily lives of the local population and, as a result of actions taken by the authorities, a component of national conscientiousness and an instrument for building social cohesion.

Monitoring 'dubious persons'

The surveillance plan for the French-Italian border was based on the very broad list of categories of suspects that it was expected to identify. Although military intelligence was apparently the government's primary objective, the police were apparently expected to monitor a very diverse range of persons. An analysis of reports, mainly from the Commissariat of Beausoleil near Monaco, suggests that spies could easily be included in the large group of people denounced as 'dubious persons'. For example, one report describes a hideout used by two recognized spies:

> Their apartment was the meeting place for a lot of somewhat dubious persons, flashy foreigners, alien and highly suspect demi-mondaines, Italian officers in disguise and also croupiers from the casino at Monte-Carlo.[3]

---

2 See Departmental Archives for the Alpes-de-Haute-Provence (ADAHP), 4 M 50: *État nominatif des facteurs qui pourront être appelés à exercer la surveillance de la frontière, des établissements militaires et des lignes de chemins de fer*, Execution of the Confidental Circular (1 May 1899); Letter to the Posts and Telegraph Department in the Department of Basses-Alpes, 6.9.1899; *Instruction pour la surveillance des frontières terrestres des établissements militaires et des voies ferrées*, 6.9.1899.

This description is similar to others made by criminology experts who, during the same period, explained that the emergence of international, cosmopolitan and imperceptible crimes warranted the development of police cooperation at the European level. As potential spies, dissimulators, crooks and delinquents, international criminals were thought to slip through the clutches of the police and avoid border controls (Reiss, 1907: 93-94; Rouché, 1909: 13-14; Locard, 1909: 392-393; Jäger, 2006). They preferred to operate in areas near the frontiers and their main advantage lay in their ability to take on new identities: without awakening the suspicions of the police, they could change name, clothes, appearance, language, and habits. They could remove all traces of their real identity and fade into the crowd. Influenced by these powerful myths, of which the most extreme example was human traffickers (known as 'white slavers'), the police reacted by extending the range of categories to such a point that all individuals were transformed into potential suspects (Chaumont, 2009). In a letter providing a definition of the types of assistance that various public services were expected to offer the security forces, the focus was "on suspicious intrigues and dubious characters", even though the definitions for these were somewhat imprecise.[4] This lack of precision was also present in the instructions given to agents in the field in which they were asked to pay attention to the appearance of individuals and the interpretation of their characteristics and subterfuges:

> In general, agents must report, using rapid forms of communication, those individuals whose appearance seems suspect, or whom they have surprised in the commission of acts covered by application of the Law of 1886 on Espionage. They must, in this regard, act with great prudence and be sure not to confuse ordinary travellers or tourists, who may display an interest in military installations, with the spy, whose goal is to obtain true intelligence.[5]

In this context, where honest folk could be taken for the most dangerous of criminals or for persons representing the greatest threat to the safety of the State, surveillance of frontiers was a natural extension of national defence activities. New categories of persons were now designated as major threats, where previously they had merely been seen as a problem for public law and order.

---

3 Departmental Archives for Alpes-Maritimes (ADAM), 4 M 1357, *Rapport du commissaire spécial de Beausoleil*, 21.3.1905.

4 AD AHP, 4 M 50, Lettre du directeur des douanes à Chambéry à M. l'Inspecteur de Briançon, "Concours aux autres services publics. Surveillance des frontières et des établissements militaires", 2.9.1899.

5 AD AHP, 4 M 50, Circulaire du Préfet aux Gardes champêtres, "Surveillance des frontières terrestres, des établissements militaires et des voies ferrées", 27.7.1899.

This militarization of security services was discussed in a book published in 1910 which analysed various customs regimes. It highlighted the efficiency of surveillance of borderlands, and in particular the French-Italian frontier, as part of the campaign against smuggling and fraud (Moye and Nogaro, 1910). The French system was thus presented as a series of defensive lines with intermediary zones that were patrolled by surveillance officers. These lines made it possible to carry out 'a preliminary filtering of fraud" which then allowed the officers to separate the 'honest transporters' from the 'cheats'. Between the two principal lines, a 'corps of customs patrols' "formed a natural battleground between customs officers and their enemies, smugglers" (Moye and Nogaro, 1910: 169-170). According to this description, the law enforcement services responsible for border surveillance constituted a body of agents including patrol members, detectives in charge of discovering clues of suspects, soldiers for planned ambushes, and informers providing information on forms of transport that might be suspect.

The frontier police took a particular interest in deserters, as shown in several episodes reported by the Special Superintendent for Menton. In April 1907, he recorded the case of a young man whose parents were Italian and living in France and who would regularly visit a café on the other side of the border. Recognized by the Italian *carabinieri* and, according to his report, "most probably denounced to them as an objector to military service", he was arrested and taken to the patrol station. He managed to escape, after an exchange of blows and injuries which ended in a comical confrontation:

> Once on French territory, he stopped and hurled insults in reply to insults and threats from the gendarme who had chased him on the other side.[6]

Even if there were few cases of this type on either side of the frontier, they demonstrate how relations between police forces had become more difficult. At the same time, the local population found themselves, against their wishes, under pressure and caught in the middle of tensions which contaminated the work of border agents implementing national security policy. Such a banal situation can suddenly release a disturbing spiral. In 1904, a botanist wrote to the Ministry of Foreign Affairs for "permission to gather herbs" along the frontier. This request launched a long exchange of letters between the Prefecture of Alpes-Maritimes and a consular agent as to whether this authorization should be given. To check his reputation, an enquiry into his morals was carried out in order to verify his *bona fides*. In particular, the report noted the sincerity of his 'republican opinions'.[7]

---

[6] AD AM, 4 M 1357, Letter dated 1.4.1907.

[7] AD AM, 4 M 1353, Exchanges of letters, January 1904.

Such an extension of the surveillance procedures gave rise to doubts within the police force itself. In a letter addressed to the Prefect, the Special Superintendent of Nice reported on the consequences of receiving erroneous information from Italian informers who were cooperating in the fight against libertarians and anarchists. Many abuses identified by malicious reports, which look more like denunciations, now had a serious impact on the lives of persons for no valid reason:

> [...] For about fifteen years, many people were subjected to a persistent and irritating surveillance or even expelled from French territory, even though their political opinions were in fact quite innocuous. The French police were thus, against its will, condoning hatred and personal vengeance by Italian informers sent to France.[8]

Checking identities

The border services' preoccupation with controlling and verifying identities is part of the long history of zones of contact and of strategic areas in terms of territorial defence and protection (Denys, 2002; Sonkajärvi, 2008; Denis, 2008). By the end of the 19th century, the emergence of new identification technologies, tighter checks on foreigners, introduction of new procedures for expulsion and the move towards national police networks contributed strongly to their enthusiasm for checking the identity of travellers (Caestecker, 2002).

Due to improvements in methods for checking identities, these gradually became an essential component in border surveillance procedures and particularly for persons travelling through areas designated as defence zones. The introduction of an integrated national identification system, as a result of the activities of the Prefecture of Police's investigation department, made it easier for local police forces to respect the system of coordinated verification of identities (About, 2004). Information about how these verification procedures, though far from complete, were carried out is available not only for most major French towns, but also for departments along the Italian border. Following a reform of commissariat filing systems launched by a series of instructions in 1903, the prefectures received several *États signalétiques*[9] with details on thousands of persons which had to be transferred into forms. Many of these documents gave very explicit information on the movement of identified, expulsed and convicted persons in departments in

8 AD AM, 4 M 1357, Letter from the Special Commissioner for Nice to the Prefect, 6.11.1909.

9 These publications listed expelled foreigners, political suspects, liberated criminals, escaped prisoners and deserters.

South-East France who were now classified for particular surveillance.[10] These *États signalétiques*, which gave details of the person's physical characteristics, made it possible to obtain precise descriptions of a large number of possible suspects among the many persons suspected of anarchy and classified in separate files for 'anarchists believed to have disappeared', 'foreign anarchists not having been expelled living outside France', 'anarchists believed to have disappeared', 'nomadic anarchists', or 'foreign anarchists expelled from France', to which were added 'deserters and objectors'.

As elsewhere, departments along the Italian frontier also received new tools from Paris which gave them access to portable extracts of files from the central archives kept by the Prefecture of Police. In 1894, several frontier commissariats in the Department of Alpes-Maritimes acknowledged reception of albums containing photographs of persons expelled or thought to be dangerous and were therefore to be subjected to special surveillance (Bertillon, 1894).[11] Designed to be used for reference, these compilations were in fact difficult to use without special training. Identities were mostly verified in the presence of the persons concerned and were increasingly compared, after the 1900s, with documents for confirming their social and civil status.

Already during the mid-1890s, several orders had already insisted on the importance of a careful examination of papers carried by foreigners crossing the frontier. Particular care was to be taken with persons wearing a military uniform and, for this purpose, the Ministry of War issued several recommendations on how to reinforce these checks.[12] These verifications became increasingly the norm and were also adopted on the Italian side, as noted by the Special Police Superintendent at La Condamine:

> [...] I have the honour of informing you that it is exact that unknown French persons crossing the Col de Larche are required to show their identity papers to Italian carabinieri, customs officers and alpini [elite mountain soldiers], failing which they are invited to return to the other side of the frontier. [emphasis in the original][13]

The severity of measures applied by Italian officers is illustrated by several examples showing what happened during individual checks. In

[10] AD AHP, 2 R 59-63AD; AD AM, 4 M 488.

[11] AD AM, 4 M 178, Avis de réception d'un courrier de la Sûreté générale, 26.9.1894, Commissariat spécial de Menton et La Turbie. This album was specifically dedicated to police forces situated in border departments.

[12] AD AHP, 4 M 50, Note de service, Ministère de la Guerre, "*Examen des papiers des étrangers qui franchissent la frontière en uniforme et par mégarde*", 12.12.1895.

[13] AD AHP, 4 M 50, Letter dated 16.8.1896.

May 1896, a grocer on a visit to his child living with its wet nurse in Italy was accompanied throughout his stay by two *carabinieri* "because he did not have a passport"; in July, three French tourists on bicycles, having crossed the frontier at Argentière, had to return as they did "not have satisfactory papers"; in the same month, a merchant from Digne was prevented from crossing the border "because he had only [...] a voter's registration card".[14]

These tight identity checks initiated a spiral that forced both countries to tighten their border controls and impose identity checks at border crossings, even though the border officers had no way of checking the information contained in the documents produced, the true details of their criminal record, or any other information obtained during checks at other border crossings. The absence of centralized information resources and of an efficient communications network explains without doubt the growing number of individual checks, which were thought to create a rampart and a deterrent for as many people as possible.

The question of identity controls quickly became a matter of the highest importance and a key issue in diplomatic relations between the two countries. A dispatch from the Sûreté Générale, dated January 1897 and sent to the Prefecture of Basses-Alpes, indicated that the Italian Ambassador in France had made an official complaint to the Minister of Foreign Affairs about the severity of checks by French border officials on Italian nationals and, in particular, new demands in terms of documents required by controllers:

> Although previously, the simple production of recognition cards was enough to satisfy the border officials, they now require that they have official certificates or passports.[15]

The official nature of these documents was expected to reassure the authorities who were now able to set new criteria for checking identity papers at the local level. These events show that identity checks at the borders attracted the attention of the highest echelons of government and that they were now a major question of public security.

The expulsion of undesirable 'nomads'

Protection of national territory was one of the most important issues to be taken into account when developing a surveillance system in border regions. It facilitated, in these areas in particular, a gradual slide towards a particularly extensive control structure, which operated not

---

[14] *Idem.*

[15] AD AHP, 4 M 50, Courrier de la Sûreté générale au Préfet des Basses-Alpes, 18.1.1895.

only at the frontier itself but also in a large area inside the departments. In this context, certain groups, such as caravans of Gypsies, were now challenged not only on their presumed status of 'nomad' but also as 'foreigners' in transit, because of questions about the legality of their civil status.

Worries about their identity and eminently suspicious nature of Gypsies gave rise, in the early 20th century and most of Europe's Mediterranean countries, to the authorities using particular zeal in checks on them. The wealth of archives on the passage of these communities and their treatment by the police constitutes, in this regard, a particularly interesting insight into the way the border region had been transformed. Gypsy communities crossing the borderlands became, at this time, one of the most targeted groups for police surveillance. At the same time, a large number of scientific reports and press articles led to the construction of a particularly negative image of the 'International Gypsy' (Donert, 2007; About, 2010): they were suspicious by definition, as specified in regulations issued by police authorities; they fell into a somewhat vague position with regard to rules on nationality as it was difficult to determine from which nation they came; and finally, for the general population, they were thought to be responsible for the lack of security in the countryside (Filhol, 2007; Asséo, 2007; About, 2012).

Towards the late 1890s, caravans travelling regularly from one side to the other of the Franco-Italian border were not subjected to any specific controls. There was even some flexibility, as described in a reference to a family of seven travelling in a caravan in April 1898. The head of the family, an "animal trainer", entered Italy at Fontan where his identity was checked by gendarmes carrying out routine identity checks. His documents were examined and the names of each member of the family were scrupulously noted. This superficial check did not lead to any further verification and they were allowed to continue their journey without any further difficulties, as can be seen in the notes recorded at the time:

> Their crossing at Fontan has not lead to any complaints. The town authorities took no measures with regard to their passage, and their stay in the commune lasted no longer than the time required for them to fulfil customs formalities.[16]

This simple note, which reports a frontier crossing, is evidence of a normal and tolerated passage in a region of France where numerous groups of gypsies worked at funfairs.

16 AD AM, 4 M 179, Procès-verbal de gendarmerie, Brigade de Fontan, "Passage à Fontan d'une bande nomade", 19.4.1898.

Shortly before 1914, radical changes in the political situation led to many itinerant communities, defined as '*grand voyageurs*' and making regular passages across borders, being challenged repeatedly by police officers who sought to forbid them entry or, on the contrary, to accelerate their expulsion. The Côte d'Azur was no stranger to these episodes which gave rise to unusual encounters between rich holiday-makers of the Belle Époque and exotic bands of 'nomads'.

Between 1913 and 1914, during a particularly harsh winter, the French authorities recorded many references to a group of Gypsies, thought to be Russians, in the South-East region and to reactions to their presence which describe how this frontier zone between France and Italy had been transformed during the previous years. In December 1913, the caravan is registered as "the Demestre group of nomads", with no reference to their professional activities. They were seen near Briançon where they tried unsuccessfully to cross into Italy by the Montgenèvre Pass. Telegrams sent to and from prefectures in the frontier departments provided regular reports indicating that their apparent objective was to cross the frontier by the coast road, which seemed to have been strongly encouraged by the French authorities. In fact, at that very moment, the *Sûreté* had sent a file with expulsion orders for ten members of this same Demestre family, some of whom had already been sentenced to leave the country in 1897.[17] Despite these antecedents and judicial decisions that would normally lead to the family's expulsion from the country, for the local authorities, the frontier was quickly becoming a solid barrier. The Police Superintendent in Menton gave a detailed account to his superior officer on conditions which made it impossible to force people to cross the frontier against their will:

> I believe that I have the obligation to inform the Prefect that the police of this town [Vintimille] will most certainly oppose their entry into Italian territory in accordance with recent and formal instructions issued by the border officers with regard to "Zingari" who are being hounded everywhere, and it is forbidden to allow them to enter the Kingdom, on pain of returning them to the frontier while the officers who allowed or tolerated their passage being responsible for all related costs.[18]

This quotation demonstrates how controls were being tightened this specific category of persons whose race, together with administrative procedures, was used to justify the adoption of extraordinary measures in both France and Italy. In another example, the Police Superintendent

---

17 AD AM, 4 M 179, Courrier de la Direction de la Sûreté générale, service des Nomades, 12.1.1914; nine explusion orders promulgated between 30.8.1897 and 17.10.1912.

18 AD AM, 4 M 179, Lettre du Commissaire Spécial de Menton, 26.12.1913.

in Menton reported on the case of two families of 14 persons named Yovanovitch who were traveling to Bosnia with direct passes for Savona and substantial financial resources and were detained at the frontier at the beginning of December 1913. Faced with resolute opposition from Italian customs officers, it was necessary to obtain a special transit permit for the family to travel, under escort, to the Austrian frontier where they could be given a temporary and limited pass to travel from one country to another.

A second group of caravans travelling through towns along the Côte d'Azur was met by a wave of protests. The Prefect for the Alpes-Maritimes received complaints from mayors who claimed the presence of this caravan in their commune led to their supposed costs. Many people wrote letters suggesting that the families be transported to Marseilles so that they could be sent away by boat or any other way of ensuring a "rapid evacuation of the tribe". However, the Ministry of the Interior nevertheless pointed out the financial and diplomatic difficulties involved in an expulsion by sea. A sombre description of the situation was published, early in January, on the front page of a local newspaper:

> A group of 30 persons, men, women and children, in three caravans drawn by four horses, travelling from Valence in short stages, arrived in Nice. They were sent to La Trinité-Victor, from where they continued their journey to La Turbie, then to Cabbé-Roquebrune and Menton in order to enter Italy.
>
> But the Italian authorities refused them passage so the "undesirables" were arrested between Menton and Roquebrune.
>
> The group had made halts in all the communes traversed by the caravans in order to gather supplies. The men, women and children had looted, stolen, threatened and spread terror everywhere; they feared no one, because how could you stop thirty persons who would only fill the prisons with vermin?[19]

These 'worrying guests', as they called, were accused of every type of crime and their presence was considered to be incompatible with public law and order. This description with details of every possible characteristic of a gypsy band, based on the usual images published by the press, drew attention to the large number of women and children, their supposed filthiness, and the impact of their passage through the communes. The very idea that they might be seeking a long-term stay provoked grim terror in observers who scattered superlatives throughout their reports.

---

[19] AD AM, 4 M 179, "Les nomades sur la Côte-d'Azur. Requête à M. le Préfet", in *Journal de la Corniche*, 11.1.1914.

Later correspondence demonstrated the feverish attitude taken by the authorities who were closely following the peregrinations of this group of caravans. In a letter sent to the Direction of the Sûreté at the Ministry of the Interior, the Prefect of Alpes-Maritimes explained that a few days earlier the group had been turned back at the Italian border and was now travelling towards the German frontier and would cross several departments along their route. A few months before the outbreak of the First World War, the tale of this group of gypsies, and their search for a crossing point despite being turned back everywhere, finally ended in Nice's hinterland.

Conclusion

In the morning of 30 August 1908, the Italian Touring club organized a meeting which would commemorate the installation of a signpost at the frontier on the road between Fontan and Cuneo. The entire delegation, of about thirty people, including the Italian Special Superintendent and Customs Officer and their French counterparts, participated in the ceremony which was accompanied by speeches and applause:

> After a glass of Vermouth, served on the parapet of the road, the ceremony broke up with cries of: Viva Italia! Vive la France![20]

This meeting, described as perfectly "friendly and cordial", was not just to commemorate the installation of a simple signpost and provoke the enthusiasm of the participants. Visible from every direction, it became a landmark, the physical and psychological manifestation of the frontier, symbolizing the limits of this and another country in the minds of the local population and of foreigners alike. It demonstrated the importance of a convivial neighbourliness, or indeed of greater closeness. Further the description of the commemoration provides another perspective, that was, in all probability, not suspected by the participants; the installation of the signpost was not only inscribed in the memory of those who attended, but also formally recorded for history by its presence in the multitude of police reports that formed the basis of the network of specialized information on border controls.

The dual significance of this episode, as a celebration of the border and a report of a public meeting at that same border, allows us to understand the profound changes occurring in borderlands that were placed under the supervision of new security services as described above. These authorities created a 'buffer area' which defined national territory and placed the entire local population under a constant and unpredicta-

[20] AD AM, 4 M 1353, Rapport du Commissaire spécial de Fontan, "*Érection d'un poteau indicateur du Touring-Club italien à la frontière franco-italienne*", 30.8.1908.

ble surveillance. This situation, which was both drawn out and intermittent, emerged because of the empirical nature of police practices and the absence of solid methods for obtaining information that would provide a better understanding of the status and position of individuals under surveillance. This would undoubtedly explain the sometimes arbitrary extension of control procedures focusing on a growing number of *suspects*. Some of these suspects, such as gypsy families, were designated as a major threat for public law and order, and paid the price for this sometimes abusive extension which required considerable public resources.

On the other hand, this surveillance network contributed to the creation of a border spirit, which was celebrated at the time by the Touring Club. Thanks to the omnipresence of security services, the marking of territory and a shared certitude that borderlands must be supervised and made secure, a new form of awareness was being consolidated. Whether as subjects or as witnesses of this surveillance, the local population became accustomed to the physical presence of borders to such a degree that it was possible to consider protection of the border as a way of defending the nation.

References

About I. 2004. "Les fondations d'un système national d'identification policière en France (1893-1914). Anthropométrie, signalements et fichiers". *Genèses*, 54: 28-52.

About I. 2010. "De la libre circulation au contrôle permanent. Les autorités françaises face aux mobilités tsiganes transfrontalières, 1860-1930" *Cultures & Conflits*, 76: 15-37.

About I. 2012. "Underclass Gypsies. An Historical Approach on Categorisation and Exclusion in France, in the Nineteenth and the Twentieth Centuries" in M. Stewart (ed.) *The Gypsy 'Menace'. Populism and the New Anti-Gypsy Politics*. London: Hurst: 95-114.

Asséo H. 2007. "Pourquoi tant de haine? L'intolérance administrative à l'égard des Tsiganes de la fin du XIXe siècle à la veille de la Deuxième Guerre mondiale". *Diasporas. Histoires et sociétés* 10: 50-67.

Baker D.N. 1978. "The surveillance of subversion in interwar France: The Carnet B in the Seine, 1922-1940". *French Historical Studies*, 10(3): 486-516.

Baycroft T., C. Grohmann, and P. Lawrence. 2001. "'Degrees of foreignness' and the construction of identity in French border regions during the inter-war period". *Contemporary European History* 10(1): 51-71.

Bayon N. 2004. "Personnels et services de surveillance de la Préfecture de police: de la constitution des dossiers de surveillance à la mise en forme du politique (1870-1900)". *Cultures & Conflits* 59: 83-98.

Becker J.-J. 1973. *Le carnet B. Les pouvoirs publics et l'antimilitarisme avant la guerre de 1914*. Paris: Klincksieck.

Bertillon A. 1894. *Album photographique des individus qui doivent faire l'objet d'une surveillance spéciale aux frontières*. Paris: Imprimerie et librairie des Chemins de fer.

Bottin M. 1988. "La militarisation de la frontière des Alpes-Maritimes, 1878-1889". In *Les Alpes-Maritimes, 1860-1914. Intégration et particularismes* Nice: Serre: 97-116.

Bouisset C. 2002. "Le découpage de la frontière franco-italienne dans les Alpes-Maritimes en 1860 et ses conséquences: entre conflits locaux et internationaux". In C. Desplat (ed.). *Frontières*. Paris: Éditions du CTHS: 167-179.

Caestecker F. 2002. "The transformation of nineteenth-century West European expulsion policy, 1880-1914", in A. Fahrmeir, O. Faron and P. Weil (eds.). *Migration Control in the North Atlantic World. The Evolution of State Practices in Europe and the United States from the French Revolution to the Inter-war Period*. New York: Berghahn: 120-137.

Chaumont J.-M. 2009. *Le mythe de la traite des blanches. Enquête sur la fabrication d'un fléau*. Paris: La Découverte.

Denis V. 2008. *Une histoire de l'identité. France, 1715-1815*. Paris: Seyssel, Champ Vallon.

Denys C. 2002. *Police et sécurité au XVIIIe siècle dans les villes de la frontière franco-belge*. Paris: L'Harmattan.

Donert C. 2007. "Der 'internationale Zigeuner' in der Tschechoslowakei. Eine transnationale Geschichte der Grenzkontrolle, 1918-1938", in C. Duhamelle, A. Kossert and B. Struck (eds.) *Grenzregionen. Ein europäischer Vergleich vom 18. bis 20. Jahrhundert*. Frankfurt am Main: Campus Verlag: 295-314.

Filhol E. 2007. "La loi de 1912 sur la circulation des 'nomades' (Tsiganes) en France". *Revue européenne des migrations internationales* 23(2): 135-158.

Fontaine L. 2003. *Pouvoir, identités et migrations dans les hautes vallées des Alpes occidentales, XVIIe-XVIIe siècle*. Grenoble: Presses universitaires de Grenoble.

Jäger J. 2006. *Verfolgung durch Verwaltung. Internationales Verbrechen und internationale Polizeikooperation, 1880-1933*. Konstanz: Universitätsverlag Konstanz.

Jauffret J. 1988. "La défense des frontières françaises et l'organisation des forces de couverture 1874-1895". *Revue Historique* 566: 359-379.

Laurent S. 2009. *Politiques de l'ombre. État, renseignement et surveillance en France*, Paris: Fayard.

Les Alpes-Maritimes et la frontière de 1860 à nos jours. Ruptures et contacts. 1992. Nice: Serre.

Locard E. 1909. *L'identification des récidivistes*. Paris: Maloine.

Mourlane S. 1999. "Migrations frontalières et engagement politique: les communistes piémontais et liguriens expulsés des Alpes-Maritimes (1922-1935)". *Cahiers de la Méditerranée* 58: 201-211.

Moye M. and B. Nogaro. 1910. *Les régimes douaniers. Législation douanière et traités de commerce*. Paris: Colin.

Ortolani M. 2008. "La frontière des Alpes-Maritimes dans le cadre des relations franco-italiennes, 1871-1914". *Recherches régionales Côte d'Azur et contrées limitrophes* 190: 45-64.

Panicacci J.-L. 1992. "La militarisation de la frontière, 1928-1940", in *Les Alpes-Maritimes et la frontière de 1860 à nos jours. Ruptures et contacts.* Nice: Serre: 75-86.

Panicacci J.-L. 2010. *L'Occupation italienne. Sud-Est de la France, juin 1940-septembre 1943.* Rennes: Presses universitaires de Rennes.

Raviola B.A. 2010. "De l'osmose à la séparation. La construction de la frontière entre la France et le Piémont-Savoie (XVIe-XVIIIe siècles)". *Cahiers de la Méditerranée* 81: 271-289.

Reiss R.A. 1907. "Un code télégraphique du portrait parlé". *Archives d'anthropologie criminelle* 22: 93-94.

Rouché J.-S. 1909. *État actuel des services d'identification judiciaire. Du choix d'un service international.* PhD Thesis in Medicine, Bordeaux: Imprimerie commerciale et industriel.

Sica E. 2011. *Italiani Brava Gente? The Italian Occupation of Southeastern France in the Second World War, 1940-1943.* PhD Thesis in History, Waterloo, University of Waterloo.

Sonkajärvi H. 2008. *Qu'est-ce qu'un étranger? Frontières et identifications à Strasbourg (1681-1789).* Strasbourg: Presses universitaires de Strasbourg.

Thoral M.-C. 2007. "Administrer la frontière: les fonctionnaires de l'Isère et la frontière franco-italienne de la Restauration à la Monarchie de Juillet". *Histoire, économie et société* 26 (1): 85-105.

Vogel M. 1997. "Une police d'État à la frontière. La direction de la police d'État de Nice (1927-1939)". *Cahiers de la Méditerranée* 55:187-202.

North African Migration under Surveillance

Between Home Country and Host Country

Sylvie MAZZELLA

In North Africa, the "Arab Spring" made 2011 a historic year, thanks to media coverage throughout the world.[1] The major political upheavals that occurred during this period had an immediate impact on the massive movement of refugees, whether within the region or to Europe.

According to statistics provided by the International Organisation for Migration (IOM) for May 2011, nearly 780,000 people fled Libya to find refuge in neighbouring countries. In Tunisia, the official figures for the same period indicate the presence of more than 380,000 refugees[2] from Libya, as a result of the armed conflict between Gaddafi's supporters and the opponents to his regime: among them, there were not only Libyans, but also foreign refugees, mainly from Somalia, Sudan, Eritrea, Ethiopia, Iraq and the Côte d'Ivoire.

Elsewhere in the Euro-Mediterranean area, there was a considerable increase in the number of illegal migrants with nearly 25,000 arriving on the Italian coastline during the months of January and February 2011, mostly Tunisians, Libyans and Sub-Saharan Africans. This sparked a debate on a proposal for reforming the Schengen Accords, which was adopted on 7 June 2012 and which included the possibility of re-establishing temporary controls on the European Union (EU)'s internal borders in "exceptional circumstances".

---

1 Major political events in 2011 led to the downfall of existing regimes in Tunisia, Egypt and Libya, and serious political uprisings in other Arab countries (Yemen and Syria in particular): President Z. Ben Ali went into exile on 14 January 2011, after a month of violent popular unrest that the Tunisians call the 'Jasmin Revolution', led to the dismantlement of the governing RCD party and election of a constitutional convention in October 2011; President M. Moubarak resigned on 11 February 2011 following an uprising by the Egyptian people; civil war in Libya, death of President M. Gaddhafi and election of the National Transition Council's provisional government in October 2011.

2 According to IOM statistics, in 2011 refugee populations fled Libya for Egypt (278,000 persons), Niger (62,000), Chad (25,000), Algeria (19,000), Italy (11,000), Sudan (2,800) and Malta (1,100): http://www.migration-crisis.com/libya/reports/view/424.

The spectacular nature of the migration crisis in 2011 can undoubtedly be seen as a singular and exceptional episode. It was in fact an important, though somewhat spasmodic, element in the migration policy followed over the last twenty years between Europe and North Africa.

To support this hypothesis, this article will review the issues involved in readmission agreements between EU Member States and North African countries as a way of controlling migration, after a discussion of dual migratory flows that are on the increase: illegal departures towards Europe and the arrival of a Sub-Saharan population in North Africa since the 1990s. In addition, it will attempt to analyse the reinforcement of migratory controls in North Africa which were set up in 2003 when the EU decided to fix its external borders through the integration of new members.

In addition to participating in the policy for cooperation over border controls and humanitarian aid, North African countries were now part of a wider – Mediterranean-wide – process as they found themselves becoming countries of destination and transit for populations from the South. Today, as work begins on drafting new constitutions, some of these countries, such as Tunisia, are reflecting on the terms of new national legislation on the rights of refugees.

North Africa: a region of emigration and transit

Traditionally, North Africa is a region of departures for Europe. More than 4 million North Africans now live abroad (nearly 2.5 million Moroccans, 1.2 million Algerians and 800,000 Tunisians) of which 85% are in Europe.

But there have been significant changes since the 1990s. The situation in terms of migration in North Africa is marked by two major factors: higher levels of illegal migration by North Africans to Europe; and a growing number of Sub-Saharan Africans in transit for Italy, Egypt, Turkey, the Balkans, or Israel, or staying for a few years in North Africa (Bredeloup and Pliez, 2005; Boubakri and Mazzella, 2005; Anteby, 2009). There are few official statistics for these migrants in North African countries: either for reasons of deliberate policy, as officially revealed by the Tunisian National Institute of Statistics in March 2011, or because existing statistics collection methods make it impossible to obtain exact figures on migratory flows (entries and exits from the country) in this region and to distinguish between migrants who are long-term residents and those who are in transit for other countries. It should be noted that nationals from members in the Arab Maghreb Union and some Sub-Saharan countries who entered Tunisia without needing a tourist visa, as authorized by the CEN-SAD[3] Agreement, are not counted as foreign residents.

Nevertheless, collective and comparative field surveys emphasize that the number of sub-Saharan migrants to North Africa and other countries along the south-eastern Mediterranean coast has increased substantially over the last 10 years: Malians, Senegalese, Ivoirians, Gabonese in Tunisia and Morocco; Malians and Nigerians in Algeria; Senegalese, Malians, Gambians and Guineans in Mauritania; Sudanese and Chadians in Libya; and Sudanese, Ethiopians, Eritreans and Somalis in Egypt.[4]

These sub-Saharan migrants can become long-term residents (Peraldi, 2011). For many years, they have provided cheap labour for North African countries. Libya, in particular, has taken advantage of this opportunity, having seen the benefits – well before Morocco, Algeria and Tunisia which have stricter labour laws – of having a labour force that could contribute to the country's new economic boom. Over the last four decades, Libya has regularly encouraged the migration of workers from black Africa (Sudan, Chad, Eritrea) (Pliez, 2003).

There are several factors behind the arrival of Southern migrants in North Africa: the lure of a "new Europe" in 1990 which created the illusion of freedom of movement, in other words that North Africa was already part of Europe; ethnic, economic and political conflicts and natural disasters which displaced populations across the African continent; cooperation policies between North Africa and the rest of Africa, encouraged over the last 15 years by countries, including Libya, as part of an African rapprochement policy (CEN-SAD) and by new agencies such as NEPAD (New Partnership for African Development).[5] South-

3 International organization with 28 African Member States, created on 4 February 1998 in Tripoli at a Heads of State Summit Meeting (Libya, Mali, Niger, Sudan, and Chad), with the aim of "guaranteeing freedom of movement by persons and capital between Member States and the freedom to live, work, acquire property and set up a business in any of the Member States".

4 The Saharan Territories Programme, as part of a Concerted Incitation Action *Espace et Territoires. Société, économie, cultures, langages, représentations* at the Ministry for Research (Project Leader: Olivier Pliez (CNRS, CEDEJ)), 2003-2006; the IRMC Programme *Le Maghreb et les nouvelles configurations migratoires internationales: mobilité et réseaux*, (Project Leader: H. Boubakri), 2000-2003; the FSP-IRMC Programme *Étudiants étrangers au Maghreb et dans l'espace euro-méditerranéen* (Project Leader: S. Mazzella), 2005-2009.

5 Regulation of inter-regional freedom of movement in Africa was a major priority for the Economic Community of West African States (ECOWAS/CEDEAO) which, on 18 January 2008, adopted a Common ECOWAS Approach to Migration with the objective of guaranteeing freedom of movement within the region, a condition *sine qua non* for encouraging economic development. At the same time, in partnership with the West African Economic and Monetary Union, ECOWAS proposed the rein-

South migration has also developed as a result of greater access of air transport and the development of major roads in Africa over the last 10 years (Choplin and Lombard, 2010).

One of the results of this dual cooperation policy has been the emergence of a multifaceted sub-Saharan migration: refugees, adventure-driven migrants, sportsmen and women, traders, students in public and private higher education, diplomats and international executives, such as the African Development Bank's 800 employees and their families now living in Tunis (Boubakri and Mazzella, 2005).

Externalization of border controls

The North African region, because it has become an area of emigration and transit, participates in many trans-border cooperation programmes since the 1990s as part of European migration policy. This policy for Euro-Mediterranean collaboration has generated a large number of agreements which are expected to stop the passage of illegal migrants: legislation covering sanctions against organizers and participants in illegal emigration networks, training of coastal patrol personnel, installation of new detection equipment,[6] trans-border projects for increasing frontier security in return for more substantial economic and cultural cooperation projects.

The Euro-Mediterranean Conference in Barcelona in November 1995 launched what is now known as the Barcelona Process and established a multilateral framework for cooperation between the EU and 12 Mediterranean countries and territories. This conference was the continuation of a first meeting held in 1990 as part of the fairly informal arrangements known as "the 5+5 Dialogue"[7] which is still active: its last summit was held in Tunisia in 2003. The Barcelona Conference led to other initiatives for cooperation under the European Neighbourhood Policy (ENP) that were later extended to cover the Southern Mediterranean,[8] and later as part of the Union for the Mediterranean.[9] The objec-

---

forcement of controls on inter-regional and inter-continental migration by creating a unique biometric, similar to the Schengen visa, for certain non-West African countries.

6 There is a massive increase in the use of new technologies for external border surveillance: detection probes for CO^2 and heart beats, thermic and infra-rouge cameras, and new generation radar for the detection of landings.

7 Group composed of Tunisia, Algeria, Libya, Morocco, Mauritania (North Africa) and Italy, France, Portugal, Spain and Malta (Europe).

8 This neighbourhood policy provides additional funding for public and private actors with support from the African Development Bank and the World Bank. It is based on a financial and operational convention: the European Neighbourhood and Partnership

tives that drove the multiplication of these political cooperation institutions were ambitious: to place the question of border security at the centre of a programme for economic development and to encourage Euro-Mediterranean cooperation.

Re-admission agreements: an instrument for trans-frontier cooperation

A re-admission agreement is a legal convention by which the signatory states agree to accept the return of their nationals after their arrest as illegal migrants in another country and, in addition, the return of foreigners who are not their nationals but who transited their territory prior to entering and being arrested in the other country. These are just one of the key instruments in the EU's policy for reducing illegal migration over the last twenty years (Charles, 2001). In return for signing bilateral re-admission agreements, the Euro-Mediterranean countries enter into negotiations for financial and technical aid, cooperation plans and quotas for legal immigrants from countries that express a real determination to combat illegal immigration. Since the mid-1990s, we have seen an increasing number and variety of cooperation programmes as part of readmission agreements between EU Member States and North African countries.

One of the first agreements between European and North African countries was signed by Spain and Morocco in 1992 which related to the situation in which Moroccan and Sub-Saharan migrants were camped outside the towns of Ceuta and Melilla. This agreement led to the annual expulsion of a thousand migrants receiving social assistance in 1993 and 1994.[10] However, it was nearly another four years before Rabat agreed to readmit 65 illegal migrants from Sub-Saharan Africa who had transited Morocco. This long delay was used to negotiate improvements in the situation of Moroccans in Spain, together with a reduction of Morocco's external debt. In 1998, Tunisia was the first North African country to

Instrument (ENPI) and one of its priorities is the Cross-border Cooperation (CBC). Between 2007 and 2013, 13 cooperation projects were signed as part of these programmes by Tunisia and Italy, with €8 million in grants.

9 The Union for the Mediterranean, created in July 2008, brings together the 27 EU Member States, the 10 countries involved in the Barcelona Process since its foundation, and Eastern European countries who are candidates for entry into the EU. Its main priorities include: civil protection, de-pollution of the Mediterranean, development of solar energy in the Mediterranean and the Euro-Mediterranean University.

10 *L'immigration irrégulière subsaharienne à travers et vers le Maroc*, ILO, 2002. The solution applied later, and still in application, is to issue expulsion orders for these migrants and allow them to cross the Moroccan border in the hope that they will then voluntarily return to their country of origin.

sign a readmission and cooperation agreement with Italy[11] which contained conditions on migration controls and patrols in territorial and international waters with the EU. This agreement focused on ensuring fair treatment for nationals legally resident in the other's territory. It also laid down conditions for cooperation on the prevention and reduction of illegal immigration: each country was obliged to readmit its nationals when they had entered or were resident illegally on its territory, and to readmit foreigners, except nationals of the Arab Maghreb Union Members States, who were arrested as illegal migrants in the territory of either party where there was proof of their having entered directly from the territory of the other party (Ben Jemia, 2005).

The Euro-Mediterranean Association Agreements[12] with EU countries also include a re-admission clause. Algeria signed four of these agreements between 1994 and 1997 (with France, Germany, Spain and Italy) for the return of its illegal migrants and of foreigners who had illegally transited its territory. In 2003, Mauritania signed a re-admission agreement with Spain for the return of its nationals and of migrants from third-party countries arrested in Spain having transited through its territory. Mauritania then sends third-party nationals to its borders with Mali and Senegal (Ahmed Salem, 2010; Omar Bâ and Choplin, 2005).

Since 1998, Italy and France, as Europe's southern border, were particularly active in developing cross-border cooperation with North Africa. These two countries have signed no less than 30 re-admission agreements (with arrangements for police training in emigration countries, development and cooperation plans, and annual quotas of visas with priority given to certain categories of workers) with Morocco, Tunisia, Algeria, Mauritania, Egypt and Turkey. In 2004, Italy played a key role in persuading the EU to lift the embargo on sales of arms by Libya in exchange for a convention banning chemical weapons in Libya and an agreement on its active participation in the fight against clandestine migration (Perrin, 2011).

EU readmission agreements have evolved into a system of predominantly bilateral readmission, involving various migration countries, whether rich or poor, democratic or totalitarian, stable or fragile. In recent years, cooperation over readmission has become one method, among others, for consolidating the framework for wider bilateral

[11] This agreement, which has not been published, came into force on 23 September 1999, is available on http://www.stranieri.it/legislazione/italia/accordi/accordi_TUV/Ab98_tn.htm.

[12] Between 1998 and 2005, the European Union negotiated Euro-Mediterranean Association Agreements with seven Southern Mediterranean countries (Algeria, Morocco, Tunisia, Israel, Lebanon, Egypt, Jordan) for the creation of structures for North-South discussions on policy and free trade.

cooperation, which covers other areas of strategic (and sometimes more important) policy such as security, energy, trade and the fight against terrorism.

Making third-party countries more responsible: the limitations

Cross-border controls can contribute to a situation where migrants become illegal because of loopholes in political and legal measures: for example, by restricting the number of asylum applications over the last ten years, since rejected asylum seekers now find themselves in an inextricable situation; by refusing to allow students on study visas to renew their visas because they have failed their exams; and by not clarifying or not making the political and legal arrangements required for the implementation of re-admission agreements. Various studies (Charles, 2001; Cassarino, 2010) emphasize that a re-admission agreement does not guarantee respect for international standards or protection of fundamental rights. The Council of Europe is now calling for negotiation of re-admission agreements only with countries that fully comply with the Declaration of Human Rights and the 1951 Geneva Convention, that establish effective procedures for asylum-seekers, and that protect the right to freedom of movement for their nationals by not introducing penalties for illegal entry and exit from their territory (Cassarino, 2010, 54).

What can be done, however, about procedures for the readmission of illegal migrants between third-party countries when there are no formal conventions between them? As readmission agreements do not exist between North African countries or between North African and Sub-Saharan countries, expulsions to neighbouring countries are governed by 'informal' national practices.

Situation of migrants and refugees in Tunisia

During the period 2003-2010, Tunisia followed Morocco's example in tightening up its legislation (Ben Jemia *et al.*, 2004): more police controls, legislation amendments (in February 2004) requiring new passports and travel documents and heavier penalties for crimes linked to trafficking of migrants which will apply to all persons having a direct or indirect link to the crime in question. Foreign migrants who entered Tunisia illegally during this period (or who have since become illegal, in the case of students with valid visas) have told harrowing tales of their delivery to the Libyan border *manu militari*.

Tunisian legislation allows African foreigners to stay in the country for three months after their visa has expired. However, Ivoirians and Malians arriving without a visa, as authorized by the legislation (unlike Cameroonians, for example), later find themselves in difficulty. If they do not have a university registration or work permit, i.e. without valid papers, they are arrested as illegal migrants. In numerous interviews carried out during this period, diplomatic staff in African Embassies (Senegal, Mali, Côte d'Ivoire, Nigeria) in Tunisia expressed concern about the confused and difficult situations in which their nationals found themselves: from repeated expulsions for overstays of valid visas (three months for tourists), arrests for forged travel documents or student cards, and attempts to emigrate illegally from Tunisia. Few migrants are able to pay for a ticket home, especially as they probably used all their savings to pay for their journey. Although the African embassies can, in some cases, repatriate them and pay the re-admission costs, most of the time they do not have the funds to offer them any assistance.

The situation is even more complicated when Africans die or are lost at sea as they must contact families and arrange for the identification and burial of bodies. Managing migration and re-admission is all the more difficult for countries with no consulate and only an honorary consul with Tunisian nationality, as is the case for Cameroon.

Creation of retention centres in 2003

In 2003, Tunisia created retention centres known as 'temporary welcome centres'. Today, they no longer exist but, until 2010, they served as police out-stations along the Tunisian coast for housing illegal migrants before being expulsed across the Libyan border. This was confirmed by IOM's representative in Tunis in 2005 who was investigating the existence of these 'informal' centres in order to convince, in vain, the Tunisian government to organize official expulsions in collaboration with IOM and the Tunisian Red Crescent.

> "There's a centre in Tunis. They don't hit you, you can use the telephone. You spend about 15 days minimum to a month there. You are better treated than in Europe, at least that's what the Nigerians I met there told me. You have a good lunch, hot water, mattresses, it's clean but once you leave … From Tunis, you go to Sousse, from Sousse to Gabès and from Gabès to Medenine. They give you food, sometimes they put you in chains. The border is at Ben Guerdane. But there are bandits there, soldiers who know that the authorities are far away. There are no officers any more, they take your money. They leave you at the border during the night, if you run quickly, you might dodge the Libyan customs officers. In the group, there were also women, we had to leave them. I walked through the night with another group, we walked as far as Zuara, and then to Surman. Our clothes were filthy. Between Zuara and Surman, there were marshes, wire fences 5 metres high. There was a tarmacked road about 1 kilometre away but we didn't go near it, we stayed in the fields, it wasn't the desert, just sandy clay. In Surman, no one tells you anything. The police don't check on you." During

the interview, he also mentioned that he was thinking about returning to Tunisia before his expulsion, because he did not want to stay in Libya.

Extract from an interview with Jacob, a 40 year old Mauritanian shopkeeper from Nouakchott, former Fishing Officer (Tunis, 13 June 2005). He was arrested without legal papers and imprisoned in one of these retention centres in 2003. He talked about his prison experience and expulsion to Libya.

The future of refugees fleeing to Tunisia from Libya after the Jasmin Revolution

In 2011, the situation changed: illegal migrants could no longer be sent back to the Libyan border. On the contrary, Southern Tunisia had to cope with an influx of refugees from Libya. Because of the political crisis in 2011, Sub-Saharan migrants were even more vulnerable. They flee Libyan work camps only to find themselves in refugee camps in Tunisia, Egypt and Algeria. Some risked their lives by boat crossings while others who remained in Libya were heavily suspected of being in the pay of Muammar Gaddafi's militia during this period. The emergency situation in South Tunisia is managed by the Tunisian Army, the UN High Commission for Refugees (UNHCR) and various other international agencies and charities such as IOM, the International Medical Corps, Caritas Middle East and North Africa (which provides material support for Libyan refugees and foreign asylum seekers living outside the camps in Tunisia (Iraqis and Ivoirians)), and NGOs such as the International Committee of the Red Cross, the International Federation of the Red Crescent, and the Danish Refugee Council.

The Choucha camp near the Tunisian-Libyan border at Ras Jdir was set up in order to take in people fleeing Libya during the 2011 war. Between January and March 2011, the most difficult period of the Libyan conflict, the camp held as many as 20,000 people and up to 4,000 more arrived every day. Libyans were rapidly moved into local towns and were accepted by the Tunisian population, but Sub-Saharan refugees were kept in Choucha.

By April 2012, a year after the war in Libya had ended, the camp held a total of 2,905 refugees who were living in Libya before December 2011, representing twenty different nationalities: Somalis (924 persons), Sudanese (910), Eritreans (456), Iraqis (292), and Ethiopians (213) (HCR, 2012). Statistics show that most of the refugees are male, aged between 18 and 60 (16% are minors), heads of family and Muslim. For most of them, their applications have been accepted, or are being processed, for transfer to a foreign country, of which the five most important being: USA (800 cases accepted in 2011 and 700 being processed in 2012), Norway (500), Sweden (200), Australia (110), and the Netherlands (42).

> This was the case of one Sudanese family among the 800 refugees who had been accepted for resettlement in the USA. Arriving in the Choucha camp in March 2011, the family group was composed of the couple with three young children, one having been born in the camp and the husband's brother with his family, and had been living in Libya for more than 10 years. The two brothers were political refugees from Sudan who had worked in Libya as teachers of agronomics and economics in a high school. While waiting for their departure for the USA, they were teaching in the school set up 6 months ago in the Choucha camp. The ex-teacher of agronomics had used the waiting period in the camp to set up small vegetable patches in front of the tents marked out with plastic bottles and exposed to the wind and sand. "In Libya, I felt as if I was standing on a high wire between two buildings. I was living in an open air prison with my family. It was impossible to leave because I was afraid of being suspected of treason by the Gaddafi government and thus putting my family in danger. I was living in fear. One of my colleagues was imprisoned for treason, just because he had used a green cloth to wipe the blackboard, green being the colour of the Libyan flag. I was summoned to the police station every month to inform the officer of any changes in our situation: a move, birth of a child, a new car or television ..."
>
> Today, he wants to live in a country which will guarantee him freedom of expression and movement (Interview in Choucha camp, 26 April 2012).

Today, the situation in the camp is even more difficult for the 240 persons, mainly young unmarried men from West Africa who, because they are not recognized as refugees, do not come under the UNHCR mandate. The Tunisian government and UNHCR are looking for a solution that will facilitate their departure and thus speed up its closure. The Tunisian government is keen to close the camp quickly in order to avoid its being turned into the second settlement area, after Lampedusa, for migrants shipwrecked in the Mediterranean: on 18 March 2012, 74 Somalis were saved at sea by the Tunisian army and transferred to the Choucha camp so that they could have access to humanitarian aid.

Current negotiations between UNHCR and the President of Tunisia and his Ministers for the Interior, Foreign Affairs, Justice and Human Rights focus on the adoption by the Tunisian government of legislation on the question of refugees and asylum. Although at the international level Tunisia is a signatory of the 1951 Geneva Convention on the status of refugees and the 1969 Organisation of African Union (OAU) Convention on the problems of refugees in Africa, there is no national legal framework on this question in Tunisia. Closing the camp is thus dependent on the provisional Tunisian government's agreement to change existing laws on asylum and according official refugee status to people in the Choucha camp whose applications for asylum have been rejected by UNHCR.

Conclusion

Controls of migrants between the European Union and North Africa are fairly similar *mutatis mutandis* to those established in countries along the EU's external borders: Southern European countries now becoming countries of destination for new waves of migrants (Italy, Spain, Portugal, Greece) and border islands (Malta and Cyprus) which are being asked to act as border guards for the EU's borders;[13] and countries situated on the Schengen borders whose participation on migratory controls is now a major issue for the EU. We have seen that the EU has introduced a double process: externalization of border controls, and obligations on countries of origin to take responsibility as countries of origin and as new countries of destination and/or transit.

It is possible to see the strategies of bilateral and multilateral cooperation proposed by the European Union as a way of reinforcing the imbalance between wealth and poverty by transferring the cost of migration by increasing pressure on poorer nations without any guarantees of a return on their involvement. In short, this cooperation policy could, at best, be seen as an avatar for a withdrawal on economic policy or, at worst, as a new episode of neo-colonialism.

We can, on the other hand, consider that the policy on Euro-Mediterranean cooperation is the indicator of change, a new orientation for North-South relations, where Southern countries are seen as having a dissymmetric economic relationship with the EU while, as political partners in their own right and partly responsible, on condition that they negotiate compensation, for their emigrant nationals and their immigrants. In other words, we can see that some assumption of responsibility on the part of Southern countries on the way in which they introduce migration policies which, for too long, were neglected despite the economic opportunities it might offer.

---

13 During an economic crisis, former countries of origin in Southern Europe (Italy, Portugal, Spain, and Greece) continue to send their unskilled labour to other European countries and further afield, with which they have links in their economic and political history, while becoming countries of destination and transit for people from Africa. Since 2004, new EU members have, to a lesser extent, followed the same pattern. The number of immigrants varies between 7% and 11% of the total population for former EU countries of destination (Germany, Belgium, the Netherlands, Austria, Great Britain, and France) and between 2% and 5% for the more recent entrants (Ritaine, 2005; Sintès, 2010).

Undoubtedly, these two interpretations are not mutually exclusive. It is not just a question of transferring the problem further south. Today, setting aside border controls on migrants and the emergency situation, North African countries, such as Tunisia, must ask themselves: to whom do we grant the right to asylum?

References

Ahmed S.Z. 2010. "La migration irrégulière de, vers et à travers la Mauritanie". *CARIM*, synthesis.

Anteby-Yemini L. 2009. "Migrations africaines et nouveaux enjeux de la frontière israélo-égyptienne". *Cultures et Conflits*. 72:75-97.

Barros L., M. Lahlou, C. Escoffier, P. Pumares, and P. Ruspini. 2002. "L'immigration irrégulière subsaharienne à travers et vers le Maroc", ILO Programme, *Cahiers de Migrations Internationales*, 54 F, Geneva, ILO.

Ben Jemia M., Boubakri H., and Chekir. 2004. *Rapport national sur les droits des travailleurs migrants (Tunisie)*, UNESCO Report, *Le droit des migrants dans les pays d'Afrique du Nord.* Tunis.

Boubakri H. and S. Mazzella. 2005. La Tunisie entre transit et immigration. Politiques migratoires et conditions d'accueil des migrants africains à Tunis. *Autrepart.* 36: 149-166.

Bredeloup S. and O. Pliez (eds.). 2005. Migrations entre les deux rives du Sahara. *Autrepart.* 36.

Cassarino J.-P. 2010. *La politique de réadmission dans l'Union Européenne*, DGPE Report (Department of Citizens' Rights and Constitutional Affairs), European Parliament.

Cambrézy L., S. Laacher, V. Lassailly-Jacob and L. Legoux (eds.). 2008. *L'asile au Sud.* Paris: La Dispute.

Charles C. 2001. *Accords de réadmission et respect des Droits de l'Homme dans les pays tiers. Bilan et perspective pour le parlement européen.* DGPE Report, European Parliament.

Faist T. 2009. *Beyond a Border: The Causes and Consequences of Contemporary Migration.* Thousand Oaks, CA: Pine Forge Press.

Goldschmidt E. 2002. "Migrants congolais en route vers l'Europe". *Les temps modernes.* 620-621:208-239.

Mazzella S. (ed.) 2009. *La mondialisation étudiante. Le Maghreb entre Nord et Sud.* Paris: Karthala / IRMC, *Hommes et sociétés* Collection.

Mazzella S. 2008. "Putting Asylum to the Test: Between Immigration Policy and Co-Development", in T. Fabre and P.S. Cassia (eds.) *Between Europe and the Mediterranean. The Challenges and the Fears*. Palgrave, Macmillan: 41-53.

Omar Bâ C. and A. Choplin. 2005. Tenter l'aventure par la Mauritanie: migration transsahariennes et recompositions urbaines. *Autrepart.* 36: 21-42.

Peraldi M. 2011. *D'une Afrique à l'autre. Migrations subsahariennes au Maroc*. Paris: Karthala.

Perrin D. 2011. "Fin de régime et migrations en Libye. Les enseignements juridiques d'un pays en feu". *L'année du Maghreb*, VII, Sahara en movement: 17-32.

Pliez O. 2003. *Villes du Sahara.* Paris: CNRS Éditions, *Espaces et milieux* Collection.

Ritaine É. (ed.) 2005. *L'Europe du Sud face à l'immigration. Politique de l'étranger.* Paris: PUF.

Sintès P. (ed.) 2010. *La raison du mouvement: territoires et réseaux de migrants albanais en Grèce*. Paris: Karthala.

Tapia (de) S. 2008. *Système migratoire euroméditerranéen.* Strasbourg: Éditions du Conseil de L'Europe.

Zolberg A.R. 1993. "Chemins de la faim, chemin de la peur". *Actes de la Recherche en Sciences Sociales*. 99: 36-42.

Introduction of Procedures for Migration Controls in Greece

Case Study in the Aegean Islands

Laurence PILLANT

Changes in migration control procedures at Greek frontiers over the last 10 years, and in particular at the frontier between Greece and Turkey, could be described as 'pluralist', a term often used in contemporary research to describe a system of organizations, actors and identities by which individuals define themselves. Here it is used to describe how territories, actors and their relationships have been transformed in order to control migration.

As this border is located at the geographical intersection of maritime and terrestrial frontiers and close to a major migration crossroads (Istanbul), it has become a preferred option for clandestine entry into Europe by migrants from more distant countries.[1] As a result, this and other geographical factors[2] have led to a substantial increase in the number of illegal crossings by migrants in the Aegean Sea during the 2000s. More particularly, the islands in the Northern Aegean Sea[3] have the second highest rate for arrests of illegal migrants entering Greece, after the Greek-Albanian frontier.[4] The number of migrants[5] arriving on these three islands increased substantially until 2009.[6]

1 Very few Turks cross the Aegean Sea to enter the EU. Most people arrested in this region are from more distant countries in Africa, Asia and the Middle East. Despite heavy press coverage of migration via the Turkish border, the Greek-Albanian frontier recorded the highest number of illegal crossings until 2010.

2 The routes taken by migrants are complex and dictated by a variety of factors. The route via the Aegean Sea is popular for several reasons: proximity to Istanbul (a crossroads of migration routes), dangers along northern routes (presence of mines from the Balkan wars), presence of American and European troops in Iraq and Afghanistan, tighter controls at other crossing points in the South-Eastern Mediterranean.

3 This research covers the three largest islands in the North Aegean Sea: Lesbos, Chios, and Samos.

4 Statistics collected during a field study together with other sources such as the Ministry of the Protection of Citizens' web site.

This border area has adapted to the passage of international migrants and, as a result, to submit to a number of transformations which will be discussed in this article. The first section describes the context of migratory flows across the Greek-Turkish border region and the various procedures for controlling this route and dissuading migrants from taking it: in particular, it will review the creation of new territories – such as detention areas – and the unprecedented multiplication of actors involved in migration control in the region. Nevertheless, is it possible that, by ending the government's exclusive responsibility for controlling the national territory, such transformations might in fact undermine the exercise of its powers at its frontiers? There are two possible, apparently contradictory, hypotheses: the introduction of more and different types of actors could lead either to a weakening of State power, or to a realignment or redeployment of government agencies. After this presentation of the main issues relating to migration controls at the Greek-Turkish border, the article will describe the implementation of security systems for handling a 'extended border area' and the interdependence of actors, be they private or public, national or European, governmental or non-governmental. Notwithstanding the fact that sovereignty of the State is paramount in managing these new elements, the introduction of plurality and its implications for migration controls at the border highlight the weaknesses that lie at the heart and at the outer edge of the procedures.

Migratory control in the Aegean Sea and the reinforcement of pluralities

Sensational language is continually used by the media and politicians when discussing migration to the European Union (EU), particularly in Southern European countries such as Italy (Maneri, 2011) and Greece (Sintès, 2009, 2011). Even though several European alternatives have been advanced to counter these arguments which tend to dramatize the presence of migrants in Greece,[7] there seems little point in seeing these reactions as uniquely European. Indeed, the complexity of Greek migra-

---

[5] This article refers to changes relating to migration control ad clandestine border crossings by migrants entering Europe without proper documents. They are neither illegal nor clandestine migrants.

[6] The number of interceptions has decreased in this region since 2009.

[7] In 2011, the EU began to review the situation in Greece and controls of migrants at its frontiers. For example, there were discussions about conditions of detention in Greece, which were strongly criticized by the European Committee for the Prevention of Torture, the activities of RABIT (a special rapid action unit), establishment of a Frontex office in Piraeus, and transposition of the European Directive on Returns in Greece.

tion control systems must first be seen within the international context for migration.

A new route for migrants: the Aegean Sea

The arrival of migrants from Africa, Asia and Middle East is not a new phenomenon, even though the numbers intercepted at the Greek-Turkish frontier, and in particular in the Aegean Sea (Pillant, 2010), increased significantly during the 2000s. Greece was a country of emigration throughout the first half of the 20th century, but already in the 1980s, this trend was reversed with a substantial rise in the number of foreigners arriving to the point where it was experiencing a 'migration carry-over effect'[8] as seen in other Mediterranean countries, such as Italy and Spain (Claude, 2002; Simon, 1986). During this period, Greece was also a major country of destination for Albanians and other 'ex-socialist South Balkan countries' (Parsanoglou, 2009; Sintès, 2010).

More recently, the growing complexity of migration routes and flows to Europe (Simon, 2002) not only made the journey much more dangerous, but also forced migrants to change their itinerary because of stricter controls and development of the Schengen zone (Clochard, 2010). It appears that an increase in 'clandestine' migration to the EU made journeys much longer because people now had to travel through Southern European countries. Because of these changes in migrants' itineraries and in the placement of crossing points, a new Mediterranean configuration emerged with the Aegean Islands becoming a popular destination for crossing into Europe during the early 2000s.

Historically speaking, the maritime frontier between Greece and Turkey has been a source of conflict between the two countries since its creation in 1922.[9] It is regularly subject to closures and military installations and this has inevitably had an impact on all the Aegean islands.[10] Even though the situation became gradually more open in the early 2000s (Sivignon, 2005), local interaction between the two countries relating to migration controls has remained relatively weak until now

---

8 Closure of frontiers and attempts to reduce migratory flows in Northern European countries during petrol crises led to changes in the direction of these flows towards Southern European countries.

9 Following defeat of the Greek Army in Asia Minor in 1922, the Treaty of Lausanne 1923 created this frontier which is a source of conflict between Greece and Turkey for several reasons (Prévélakis, 1997; Pazarci, 1986).

10 By the late 20th century, the North Aegean islands had become once more very attractive for Greeks as a result of government decentralization programmes. The local population had decreased substantially because of the conflicts between Greece and Turkey, the definition of the frontier between the two countries, the rupture with the Anatolian coast and high levels of emigration (Darques and Kolodny, 2004).

(Pillant, 2009).[11] Migrants usually make the short passage between the Anatolian coast and the Greek islands in makeshift and often overloaded boats. Paradoxically, in this area where the maritime limits remain a source of political conflict, the two border agencies work side by side when intercepting and rescuing migrants at sea. Greece's entry into the Schengen zone has added an extra dimension to the Greek-Turkish 'dyad'[12] and challenges the role of the State as principal actor in controlling Greek territory. As can be seen from the way in which controls on this border, and more specifically detention centres, have been installed, a wider range of actors and systems is now emerging.

Plurality of detention centres: local solutions make up for the lack of legislative provisions

Under the Schengen Agreement, crossing a frontier anywhere other than at an official border post is an offence. As the penalty for this offence is expulsion, offenders are placed in detention centres while their situation is reviewed and a decision taken on whether they should be sent back, granted asylum or ordered to leave the territory within 30 days. Although migrants are systematically placed in detention centres when they are intercepted in mainland Greece, none of the procedures currently used in the Aegean Sea take into account the need to detain hundreds of persons. The absence of a real legislative framework seems to be the primary reason for the wide variety of solutions for the detention of migrants in this region. Without any regulations covering this situation, the authorities made use of measures available under two confidential emergency plans, Poseidonio and Balkano, until 2011. These plans specify that the prefectures,[13] and more particularly the sections responsible for Civil Emergency Planning,[14] are responsible for handling the massive arrival of people during a major crisis, such as a natural disaster, although no specific details are given as to what

[11] Cultural exchanges have multiplied locally, thanks to Greek-Turkish friendship associations, in contrast with hostile attitudes and limited contacts between the two governments.

[12] According to Michel Foucher (1991), a 'dyad' is 'a section of shared frontier between two countries'.

[13] The Greek prefectures (Νομαρχια) do not have the same position as French prefectures, except in terms of their importance. Prefects are elected and are not directly involved in decisions on questions of security, which are handled by the police. The Kallikratis reforms, introduced in 2011, will change this system.

[14] Literally Πολιτικής Σχεδίασης Έκτακτης Ανάγκης (ΠΣΕΑ – Civil Emergency Planning). This agency is covered by a confidentiality clause and is responsible for the management of emergency situations, such as natural disasters, in peacetime.

arrangements must be made for them.[15] No national directives appear to have been issued for the management of such situations during this period,[16] although the Ministry of the Interior does reimburse expenses related to the establishment of detention centres. The absence of appropriate national regulations has led to a variety of local improvised solutions in border regions which result in notable differences in the cost of detention. For example, there are marked differences in the ratio between number of arrests and costs incurred. In 2009,[17] the three islands that make up the Aegean border (Chios, Samos and Lesbos) received 75% of reimbursements, even though only 34% of Greek border interceptions took place in this region. Lack of legislation and the total absence of ministerial audits of local costs[18] have resulted in differing approaches to the management and funding of detention centres.

On these three islands, the requisitioned buildings were not built for the detention of migrants and few improvements have been made by the authorities in order to fulfil this purpose.[19] Despite these improvisations, some common issues have emerged. For example, the centres were generally in a poor state of repair and lacked basic equipment, because of the decision to use of abandoned or under-exploited buildings. This also explains the diversity of 'available' resources[20] from one island to the next: a tobacco factory, a former police station, and a former prison.[21] Even though this improvisation resulted in a variety of buildings, this alone cannot explain why local authorities introduced different practices for handling the detention of migrants. To a certain extent, these practices provide insight into the attitude of the authorities towards migrants, viewed as either a nuisance or a threat. Indeed, a number of

15 The Balkano and Poseidonio plans are confidential. We were not allowed access to them and had to rely on descriptions of their contents from various actors in prefectures and at the Ministry of the Interior.

16 Despite interviews with ministerial staff (including persons at the highest levels) and research in the archives, we have found no confirmation of this.

17 Reimbursements of these costs are often late and sometimes by up to a year. Payments in 2009 probably refer to spending in 2008.

18 Ministerial spending reviews are only carried out from a distance. Some requests for reimbursements are refused as unacceptable, but there appears to be no ministerial service responsible for carrying out on-site audits.

19 We found very little information on contracts with suppliers for the early 2000s and very few services appear to have been provided and do not appear to have succeeded in rendering the buildings habitable.

20 Other possibilities were sometimes available but not used because of protcsts from the population which led to the emergence of other difficulties at the local level.

21 The tobacco factory on Chios was just a hangar with difficult access because of its distance from the island capital; the former prison on Lesbos stands above the town of Mytilini; and the former police station is situated in the centre of Vathy, Samos' main town.

abuses have been observed – for example, boarded-up windows – that cannot be attributed merely to a lack of resources. The inhuman conditions in these buildings are a source of tension and have led to the creation of collectives[22] to defend migrants' rights. Again, the strategies and methods used in the power struggle with the authorities vary on the islands. Some of these collectives offer professional services in the detention centres where they have an opportunity to play an important role but little scope for denouncing the abuses they observe. Other collectives, however, have only sporadic access to the centres and have a single option for confronting the authorities: they have greater freedom for denouncing abuse, but only limited access to information. Much of the tension that occurred between actors during the 2000s was generated by questions relating to respect for human rights and arbitrary decisions by police which prevent collectives from having regular access to the centres. The results of this power struggle accentuated the differences in treatment from one centre to another in the Aegean islands. Despite public statements and measures that sought to ensure similar treatment and to regulate practices, differences still persist in the second series of detention centres. Practices introduced in response to an emergency situation remain in place and few improvements have been made to conditions inside detention centres.[23]

Although migrants are systemically detained after interception, detention centres do not appear to receive the same resources as border controls over the last decade. A similar situation can be seen at the European level: as more efforts are made to supervise borders, less attention is paid to the basic rights and freedoms of individuals (Jeandesboz, 2008).

Migration controls and a plethora of agents in the Aegean Sea

In the Aegean islands, the army is responsible for defending the country from external threats, and in particular from Turkey. With regard to the detection of clandestine border crossings, responsibility for the islands is divided between the national coastguard and police services. With the growing numbers of migrants, preparing their officers to adapt to this situation is limited to training programmes.[24] For example,

---

[22] The term 'collective' is used here for citizens groups that are not necessarily registered 'associations' and therefore not officially recognized.

[23] Even though two centres have been built on Samos and Chios, the island of Lesbos is trying to rehabilitate a warehouse in a commercial zone.

[24] here has been no real increase in staff but reinforcements are sent to handle specific events: e.g. in August 2009, additional officers were sent from Athens to handle demonstrations during the annual meeting of NoBorder, a European anti-racism network.

police officers have been trained in using the Schengen information system and in detecting false papers. However, the most important change in terms of border controls is undoubtedly the presence of a new inter-governmental agency, Frontex.[25] Since its creation in 2004 and implementation on 1 May 2005, Frontex has become one of the major actors in migration control at European borders. For the Aegean region, it became part of the Poseidon programme in 2007 and its activities in Greece have increased since 2010 when Frontex's central operations office for the Eastern Mediterranean at Piraeus opened and an intervention force (RABIT) was deployed along the Greek-Turkish border. Even though Frontex is the principal instance for new European procedures on migration control, its link with Greece pre-dates its first operations in the Aegean Sea. Our study of the modalities of Frontex's presence in Greece showed that a precursor, the EKTHAS office,[26] was already operational in 2005. Frontex is a multi-level agency with its European headquarters in Warsaw and a national office for cooperation agreements between EU Member States. Its local office members are Greek liaison officers[27] who act as its representatives in the islands' port authorities. It is not exclusively European but operates at several levels together with other national and international actors.[28] As a result, there is a certain amount of opacity in its structure which makes its operations somewhat difficult to understand, as can be seen from the many challenges to the activities it coordinates (Rodier, 2011; Jeandesboz, 2012). It has three primary responsibilities: monitoring, dissuasion and searches. In addition, private operators are also involving in carrying out migration controls, on an ad hoc basis,[29] at the border and inside national territory: for example, travel agencies and airline companies.

The variety and complexity of actors involved in Greek migration controls are signs of a markedly different approach to the European

25 Frontex is the European agency for the "management of operational cooperation at the external borders of EU Member States" (Council Regulation (EC) 2007/2004).

26 The EKTHAS office, established in the 2000s, is responsible for coordinating "arrival controls in the ports and/or the coast of the European Union, joint patrols at sea, exchange of officers, standardization of technical methods of surveillance and monitoring of ship movements", terms similar to those for the European agency, Council of Europe, Note No. 13918/03.

27 These Frontex agents are Greek coastguards on secondment.

28 Frontex is responsible for training police officers in detecting false papers at official frontier crossings. In terms of control and dissuasion strategies, it is also required to identify current migration routes and anticipate future changes. Frontex' actions and officers have a larger brief, including arrangements for translations during police interviews of migrants and investigations into migrants' nationalities.

29 Private companies are involved in controlling identities, not in technical and technological aspects of control procedures.

position on migration and migration controls. It seems that, by adapting to new and ever-changing migration flows, controls have to become more flexible and require a wider range of actors and control points. We must now see how these various elements are linked together in order to obtain a better understanding of the impact of these procedures.

Procedures in operation: Extended border areas and fragmented controls

While social science research has focused on the State as the actor with the greatest responsibility for transforming contemporary society, we should not conclude too quickly that it is in decline, since changes in State management of national territory do not necessarily mean that it is no longer present, as the factors discussed above demonstrate. Rather, the key issues are the extensions of the border area and the emergence of control systems. In addition, based on our observations of these procedures, we should ask why these controls have become fragmented, rather than whether the State, as prime actor, has become more fragile.

Extended border areas: the result of introducing new migration control procedures

In recent years, controls of identity were no longer limited to border crossing points, apparently as part of a general move towards the criminalization of migration (Palidda, 1999), which made it possible to introduce more controls and a wider range of targets (Tsoukala, 2002) and control points. It could be argued that these controls were needed to deter irregular entry and to ensure greater flexibility in countering new routes and strategies adopted by migrants (Bigo, 1997). For control actors, the key objective was to find new ways of coming into contact with migrants. At the European level, this means 'externalization' of border operations (Audebert and Robin, 2009; Morice and Rodier, 2010; Weber, 2010; Migreurop, 2011). Similarly, at the Member States level, controls now take place within the national territory and the Aegean Islands obviously have to play their part in this process. Now identities were to be checked not only at the frontier or at official border crossings, but also throughout the country.

Given their geographical position and specific characteristics, the Aegean Islands are for the most part transit areas. Migrants intercepted at sea or on their arrival on the islands are almost always sent to detention centres. The period of their stay depends on such criteria as nationality and the specific context on each island.[30] They are usually released with an order to leave the country within 30 days. However, although travel to Athens does not involve crossing a national frontier, they are required to present identity papers when buying a ticket for the ferry. If they do not have legal papers, they must present the expulsion order which allows them to obtain a ticket for Athens. Therefore, undocumented migrants who have not been detained have difficulty leaving the island for the mainland. A circular for the Protection of Citizens issued by the Ministry in January 2010 clearly indicates that travel agencies are now responsible for checking the identities of people purchasing ferry tickets.[31] The main outcome of this circular is to make it 'quasi-obligatory'[32] for migrants to spend time in a detention centre if they want to continue their journey, usually to Athens. Thus, detention centres become another form of obligatory border crossing for entry into a foreign country, for certain categories of migrants, thus introducing an extension of the national frontier inside the country itself and making it necessary to develop new strategies of avoidance.[33]

According to this circular, travel agencies selling tickets have in fact become agents for migration control within the country. These controls apply to Greek nationals who must also present their identity cards when buying an internal travel ticket. Here we can see that detention centres, travel agencies and transport offices are now responsible for one of the most important aspects of border security: filtering. Is not this move towards extended border areas, as seen elsewhere in Europe (Clochard, 2007; Cuttitta, 2007), now the most visible feature of migration control systems? Indeed, the identification of various disparate elements – detention centres, actors, legislation – and their interaction in order to control migration seem to meet the criteria for Foucault's 'system'.[34] The extended border area is thus a territorial structure which relies on the installation of a migratory control system linking the various places where controls take place, regardless of the wide diversity of actors. By

30 E.g. the NoBorder protests on Lesbos in 2009 which led to the release of dozens of migrants from the detention centres. This demonstrates the influence that the context in the islands can have.

31 http://www.minocp.gov.gr/index.php?option=ozo_content&perform=view&id=3440&Itemid=461&lang.

32 Although many of the stratagems used by migrants relate to avoidance of identity checks (Diminescu, 2001), here migrants who have not been intercepted have to find a way to buy tickets without papers and this is much more complicated.

33 E.g. buying a ferry ticket on the black market.

34 Foucault defines this system as a "collection of completely heterogeneous elements" which responds, in the first instance, to an emergency situation with the "prevalence of a strategic objective": here it is applied to migration controls (Foucault, 2001, 299).

issuing various instruments (decrees, laws, circulars), the State has set up a system in which several agents can exercise official authority by controlling identities. In allocating these new tasks, has the government delegated its powers in order to withdraw from its responsibilities in the future?

The illusion of sub-contracting state responsibility for migration control

As we have seen above, the involvement of new actors and territories in migration controls goes hand-in-hand with an extended border area created by the establishment of a system that is supported to come into contact with the phenomenon they are expected to control. But is the State still the principal actor? In fact, the participation of non-governmental and supra-governmental actors does not necessarily imply delegation of sovereignty: rather, the government has authorized, or obliged, new actors to carry out identity controls. Travel agencies, for example, do not replace State border controls but carry out additional controls for the State at no additional cost. Having police officers or coastguards supervise the sale of tickets would be expensive, whereas a circular does not imply State intervention, yet makes provision for new forms of internal controls.

What appears to be delegation of authority or allocation of state operations to private actors via sub-contracts is in fact an accumulation of new areas and new types of control that did not exist before. For example, asking private companies offering aerial archaeological services to watch out for boats crossing in the Aegean Sea illegally[35] does not mean they take the place of other state agencies previously responsible for migration controls. It merely introduces a new element, a combination of aerial and maritime controls. Similarly, Frontex agents interview migrants in detention centres in order to gather information on their journey and passage across the border. They are not replacing the police in this task but, clearly, these interviews provide useful information that was not previously obtained during local police investigations.[36]

As the European agency for migration controls, Frontex is not expected to replace State authorities, but rather to enhance their efficiency in carrying out national migration controls and managing the external frontiers of Europe's shared space. Thus, the government takes a positive attitude to Frontex' presence in Greece, and in particular the Aegean

---

35 As occurred during an exercise carried out by Frontex on Lesbos in 2009.

36 These interviews are regularly analysed and compared with data obtained from intercepted migrants. Based on this information, state authorities can set up new strategies for controls.

Sea, and this is not perceived as a challenge to national sovereignty. Indeed, the agency's structure has been established, together with its principal intervention team (mostly coastguard officers from other Member States), as a way of providing assistance and partnership, and therefore not to conflict with existing Greek procedures. Nevertheless, the introduction of changes in the way the government carries out controls in its territory is bound to have major consequences. The absence of delegation of national sovereignty and the lack of clarity seems to have undermined these control procedures. Closer examination of relationships between these new actors at the local level allows us to identify the reasons why controls have been weakened through the installation of structures and extended borders areas without real delegation of authority.

The paradox of non-delegation of sovereignty: fragmentation of control procedures

Paradoxically, the absence of real delegation of state authority and the maintenance of national sovereignty with regard to migration, through the introduction of new procedures and the multiplication of actors, have produced two important results. First, the involvement of more actors has resulted in more opportunities for conflict and cooperation at the local level, rather than harmonized practices. The possibilities for interaction are as numerous as the current situations which in turn vary in space and time according to the specificities of the area in question. Frontex offers a particularly interesting example for this discussion. Its relations with the national coastguard and police services seem to be based on earlier links and conflicts with the civil society (Pillant, 2010). In the Aegean islands, where the coastguards have sole responsibility for maritime controls, Frontex is obliged to work with them, but this does not mean that there are no clashes.[37] Where there are only limited relations between the coastguards and the police, cooperation between Frontex and the police also suffers. Relations between the islands' control agencies may thus have a direct impact on their cooperation with Frontex. In addition, the lack of a permanent Frontex presence and the involvement of different teams with members from several European countries may not contribute to the establishment of strong links with local agencies. The arrival of a new – European – actor in these circumstances does not guarantee a unified approach to border controls in the Aegean islands, but rather it makes the situation more

37 On several occasions, Frontex agents have been described as unfriendly and uncommunicative. 'Forced' cooperation, in particular in terms of the agency's legal responsibilities in the country – a Greek coastguard on each Frontex boat – may explain the difficulties in establishing good relationships among various actors.

complex. By creating several layers of control agents without transferring real power to any of them, the government has created a situation where relations between actors become conflictual and may even affect or interfere with the main objective: control of migrants.

Second, non-delegation of powers to intermediary actors may have an impact on relations with the private sector. By asking travel agencies and ferry companies to check travellers' identities, the government is able to maintain its control over the whole country without having to allocate additional resources. Initially, this new responsibility could paradoxically compromise migration controls, since these actors are often reluctant to take on additional responsibilities if they are not paid. Under the global restructuring programme for Greek ports,[38] maritime transport companies will be responsible for checking passengers' identities before boarding. This move is not popular because the procedures are time-consuming, create lengthy delays, slow down maritime traffic, and ultimately could lead to financial loss. Such demands from governments may generate conflicts and resistance, especially from shipping companies, a very powerful industrial sector in Greece.

In addition, broadening the range of migration controls without delegation of powers or provision of additional funds or grants has led to the creation of marginal agents and their territories. Shipping companies and travel agencies, for example, operate as businesses with commercial targets and, by the very nature of their activities, are required to act as the 'last link in the chain' for migration controls. In addition, these 'last link' controls must conform to certain procedural restrictions which are obligatory for government agencies, but cannot be imposed on commercial actors. This last condition may lead to the development of new illegal practices and thus undermine further the implementation of migration controls that used to be the exclusive prerogative of the state and its agencies.

Controls on migration and closing of frontiers in Europe during the 1970s had a serious impact on the routes taken by migrants coming into this region clandestinely. Today, migration controls are an official priority in all EU countries. Given the new situation in the Mediterranean, the Greek-Turkish frontier has now become a key area for border crossings by migrants who do not have the right to enter Europe legally. Even though most European Member States have established migration

---

38 Amended in order to conform to new international, European and national codes on maritime safety and infrastructure, these reforms cover principally the Maritime and Port Code (ISPS) and constitute an additional element for migration controls in European ports (Migreurop, 2011).

policies and legislation covering specific places and actors, in fact the situation remains extraordinarily diverse.

These control structures, though composed of numerous elements, are linked together as an inherent part of the extended border. In the initial concept for the Schengen Agreement, freedom of movement throughout the European Union was expected to remove internal borders between EU Member States and strengthen frontiers with non-European countries. Paradoxically, more recent policies on migration in Europe appear to have reintroduced controls inside the Schengen zone. Although there is progress in the development of a common migration policy, governments continue to play a major role in setting up identity checks at national frontiers and in maintaining national sovereignty in terms of immigration and citizenship. The State is a key actor in the negotiation of readmission agreements with non-European countries, as with Turkey in 2001. However, even though to date this agreement has been only partially implemented, these negotiations were in fact easier than the current European Union-Turkey discussions in this domain.[39]

Government agencies still play a predominant role in the establishment of detention centres and migration control practices. In the absence of a national policy on confinement of migrants, local solutions have to be found and here again, state agencies continue to take the primary role. Even if this has not weakened the position of the State, the situation has nevertheless changed. The exercise of governmental responsibilities has been transformed by the introduction of new migration controls and the arrival of intermediary actors. These changes are not limited to migration controls, but appear to be the outcome of a more global transformation of the State itself. We are seeing a new combination of agencies and levels of decision-making that has changed the territory. And nowhere is this more obvious than when it is the State itself that introduces these changes. By creating marginal areas, it introduces new agents – political (Frontex) or commercial (travel agents and shipping companies) – and territories in which they operate, without actually delegating its authority and this has obviously had an impact on the situation. Whether or not this diminishes the key role played by the State, the State has developed a multiple presence on its territory, creating new connections with actors at various levels and new areas for controlling migration. Even though the complexity of these changes goes way beyond the question of migration controls, it offers a particularly interesting vantage point for observing them.

39 The first steps towards the signature of a readmission agreement between Turkey and the European Union took place on 21 June 2012, but negotiations have stalled because of Turkey's demands for easier access to Europe for its citizens.

References

Audebert C. and N. Robin. 2009. "L'externalisation des frontières des 'Nords' dans les eaux des 'Suds'". *Cultures & Conflits*, 73(1): 35-51.

Bigo D. 1997. "La recherche proactive et la gestion du risqué". *Déviance et société*, 21(4): 423-429.

Claude G. 2002. *Migrations en Méditerranée.* Paris: Ellipses.

Clochard O. 2010. "Le contrôle des flux migratoires aux frontières de l'Union européenne s'oriente vers une disposition de plus en plus réticulaire". *Carnets de géographes* (1): 18.

Clochard O. 2007. *Le jeu des frontières dans l'accès au statut de réfugié: une géographie européenne d'asile et d'immigration.* PhD thesis in Geography, Université de Poitiers.

Cuttitta P. 2007. "Le monde-frontière. Le contrôle de l'immigration dans l'espace globalise". *Cultures & Conflits*, 68(winter): 61-84.

Darques R. and E. Kolodny. 2004. "Turcs, Grecs et réfugiés dans l'île de Lesbos au XXe siècle" *Méditerranée* (3-4): 65-74.

Diminescu D. 2001. "Le 'système D' contre les frontières informatiques". *Hommes et migrations*, 1230 (March-April): 28-33.

Foucault M. 2001. *Dits et écrits: 1976-1988*, Paris: Gallimard.

Foucher M. 1991. *Fronts et frontières: un tour du monde géopolitique.* Paris: Fayard.

Jeandesboz J. 2012. "Contrôles aux frontières de l'Europe. Frontex et l'espace Schengen". *La Vie des idées* 10.1.2012, http://www.laviedesidees.fr/Controles-aux-frontieres-de-l.html.

Jeandesboz J. 2008. "Reinforcing the surveillance of the EU borders. The future development of FRONTEX and EUROSUR". Research Paper No. 11, *Challenge Liberty and Security.*

Maneri M. 2011. "Les médias et la guerre aux migrations", in Palidda S. (ed.) *Migrations critiques: Repenser les migrations comme mobilités humaines en Europe.* Paris: Karthala: 85-107.

Migeurop. 2011. *Aux bords de l'Europe: l'externalisation des contrôles migratoires.* Paris, Migeurop.

Morice A. and C. Rodier. 2010. "Comment l'Union européenne enferme ses voisins? Les contrôles migratoires sous-traités aux pays extérieurs". *Le Monde diplomatique*, June 2010: 16-17.

Palidda S. 1999. "La criminalisation des migrants". *Actes de la recherche en sciences sociales*, 129 (September): 39-49.

Parsanoglou D. 2009. *Grèce, pays d'immigration: Perspectives historiques et sociologiques*. PhD thesis in Sociology, EHESS.

Pazarci H. 1986. "Le contentieux gréco-turc en Mer Egée". *Cahiers d'études sur la Méditerranée orientale et le monde turco-iranien*, May (2/3): 67-85.

Pérouse J.-F. 2002. "Migrations, circulations et mobilités internationales à Istanbul" in Dumont P., J.-F. Pérouse, S. de Tapia, S. Akgönül. 2002. *Migrations et mobilités internationales: la plate-forme turque*, Les dossiers de l'IFEA 13: 9-29.

Pillant L. 2009. *Construction, évolutions d'une frontière de l'Union Européenne: L'exemple de l'île de Lesbos.* Masters 1 thesis, Université de Provence.

Pillant L. 2010. *Les transformations frontalières aux portes de l'Union Européenne: étude de cas des îles de Lesbos, Chios et Samos.* Masters 2 thesis, Université de Provence.

Prévélakis G. 1997. *Géopolitique de la Grèce.* Paris: Complexe, Géopolitiques des États du monde.

Rodier C. 2011. "Frontex, la petite muette". *Vacarme* 55 http://www.vacarme.org/article2015.html.

Simon G. 1986. "La nouvelle donne migratoire en Europe du Sud". *Revue européenne de migrations internationales*, 2(1): 9-35.

Simon G. 2002. "Penser globalement les migrations". *Projet* 272; 37-45.

Sintès P. 2011. "Manœuvres politiques et discours de l'urgence: la frontière égéenne de la Grèce au cœur d'enjeux nationaux et européens", in N. Ribas-Mateos (ed.), *El Rio Bravo Mediterráneo. Las Regiones Fronterizas en la época de la Globalización.* Barcelona: Bellaterra: 375-390.

Sintès P. 2010. *La raison du mouvement. Territoires et réseaux de migrants albanais en Grèce.* Paris: Karthala.

Sintès P. 2009. "Déjouer les fantasmes autour de l'immigration en Grèce". *Libération*, 18.8.2009: 23.

Sivignon M. 2005. "La Grèce devant l'adhésion de la Turquie". *Hérodote* 3(118); 82-106.

Tsoukala A. 2002. "Le traitement médiatique de la criminalité étrangère en Europe". *Déviance et Société* 26(1): 61-82.

Weber S. 2010. "D'un rideau de fer à l'autre: Schengen et la discrimination dans l'accès à la mobilité migratoire". *Géocarrefour* 84(3): 163-171.

Part II

Mobility and the Border Economy

Introduction

Cédric PARIZOT

In this second section, we will review border economies in the Mediterranean: the articles presented here analyze the circumventions and reappropriations developed by mobile border populations by taking advantage of loopholes and opportunities made possible by border control systems. Our objective is to assess how these reappropriations benefit from the installation of these systems.

It is now accepted that greater security at borders stimulates the development of new economic activities based on avoidance of surveillance systems and regulations (Andreas, 2001). Tightening controls forces people who live near the border or who must cross it regularly to change their habits, activities, journeys and border passages. Due to their limited resources, they are often forced to call for assistance from individuals and groups who specialize in avoiding physical obstacles (walls, barriers, etc.), surveillance systems (radar, drones, bio-metric systems) and state regulations (visas, travel permit systems, work contracts, etc.). Systems of constraints imposed on a border and mobile population always create new opportunities for smugglers of people and goods, producers of false papers, recruitment agencies specializing in foreign workers, etc. By taking advantage of this demand and of weaknesses in the systems, these actors contribute to the emergence of a complex social economy.

The border economy operates in both the formal and informal sector and plays a key role in regulating mobility. Increasing the number of controls encourages the development of smuggling and criminal activities (Andreas, 2001) and also benefits a network of enterprises already in the market. Ruben Hernandez León (2008) has shown that the US-Mexico border, by facilitating the circulation of migrants, the network of legal companies (transport, communications, banks, recruitment agencies, etc.) focusing on services for migrants has created a 'passage industry' that works *de facto* with and around control mechanisms for regulating migration. Smugglers have also imposed themselves as new 'authorities', as demonstrated in Janet Roitman's research in Central Africa (2005). Even though they operate at the state's periphery and in the informal market, they are also actively involved in regulating the circulation of people, goods and capital.

The circumvention economy also has an impact on the operations of control systems set up by states and their partners. In the early 21st century, states are not alone in controlling borders (Brown, 2009). The previous articles have demonstrated how migration policies and Schengen border control systems lead to complex cooperative efforts between these states and a multitude of infra- and supra-national public and private actors. Confronted by reappropriations and adaptations used by mobile populations to avoid surveillance systems, these public and private actors have been forced to readjust their control systems. Yet, they do not always have the opportunity or the means. Integrated and highly sophisticated surveillance systems are often less flexible and cannot be adapted as easily as the strategies of smugglers. In addition, these systems are never impermeable: there are never enough financial and human resources to maintain them and keep them fully operational. Finally, given the economic stakes involved in these activities, it is an illusion to think that all obstacles will be removed. The authorities responsible for the border thus have no other choice but to instrumentalize trespassers in order to maintain some control over their activities. Indeed, they are all the more motivated to do so because it gives them access to additional sources of funding, forms of sub-contracts and informal control mechanisms. Thus, official authorities increasingly integrate these formal and informal actors into the functioning of their control systems.

To observe the interaction amongst these different groups of actors, we have to bypass institutional and State-centred approaches. Thus, taking anthropological, socio-economic and geographical perspectives, the articles presented here provide a micro-sociological study of border economy activities. Four case studies carried out in three Mediterranean regions illustrate this process: trafficking of Palestinian workers between the West Bank and Israel (C. Parizot), management of clandestine immigrants on the island of Lampedusa (H. Friese), the international domestic service market in Lebanon (A. Dahdah) and finally, the Palestinian olive industry under Israel's confinement policy (A. Garcette).

These examples show firstly that opportunistic border economies take advantage of other forms of mobility: movements to and fro across borders (C. Parizot), migrants in transit (H. Friese), migrants well-established in their country of destination (A. Dahdah) and finally commercial interactions (A. Garcette). In these business sectors, we see a wide variety of borders and institutional contexts. It shows to what extent these actors can be inventive and adaptable. As Heidrun Friese demonstrates with the situation in Lampedusa, the borders of the

Schengen zone are managed through collaborative efforts by the Italian government, the EU, and numerous non-governmental organizations. As a result, methods of reappropriation by North African smugglers and by local actors who take responsibility for reception camps are particularly complex. Even though the situation in Lebanon offers an example of a more exclusive state sovereignty over the entry of persons into its territory, Assaf Dahdah's research shows that border economies adapt perfectly by taking advantage of opportunities generated by global trade. In a world market for domestic servants, he emphasizes the stakes involved when recruitment agencies play on discrepancies and legal loopholes in international relations between countries separated by thousands of kilometres. He also shows how the state profits from this border economy. Finally, the Israel-Palestine region offers a particularly interesting case. With the absence of internationally recognized borders, there is now a proliferation of borders and checkpoints that has led to controls being scattered over a vast territory. Here the border economy draws on highly complex forms of organization, coordination and innovation, either for avoiding extremely sophisticated surveillance systems, such as the West Bank Wall (C. Parizot), or for juggling with status and with control mechanisms on products (A. Garcette).

In conclusion, by taking into account the informal interactions between actors at the border, these articles provide us with an opportunity to deepen the debate offered in the first section on the redeployment of the exercise of state power in a globalized Mediterranean. They provide insights into how maintenance of controls on trans-border mobility continues to play with subtle interactions between, on the one hand, authorities responsible for the border (national, supra- and infra-national, public and private) and, on the other, these informal actors.

References

Andreas P. 2001. *Border Games: Policing the US-Mexico Divide*, Ithaca: Cornell University Press.

Brown W. 2009. *Murs: les murs de séparation et le déclin de la souveraineté étatique*. Paris: Les Prairies ordinaires.

Hernandez-Leon R. 2008. *Metropolitan Migrants*. Berkeley: UCP.

Roitman J. 2005. *Fiscal Disobedience: An Anthropology of Economic Regulation in Central Africa*. Princeton: Princeton University Press.

An Undocumented Economy of Control[1]

Workers, Smugglers and State Authorities in Southern Israel/Palestine

Cédric PARIZOT

This chapter focuses on informal crossings of undocumented Palestinian workers from the southern West Bank to Israel. Relying on ethnographic data, it studies the modes of organization and the ramifications of this border crossing economy throughout the post-Intifada period (2007-2010). It highlights the extent to which Israeli state authorities have *de facto* managed to involve informal actors in controls of Palestinian mobility.

While Palestinian employment in Israel has been extensively documented, little attention has been paid to the tactics involved in illegal passages into Israel. In the 1980s and 1990s, research has mostly focused on the changing role of the Palestinian population within the Israeli economy. Scholars have described how Palestinian workers, seen as a highly desirable, cheap, flexible and mobile labour force from the 1970s until the end of the 1980s (Portugali, 1989), were increasingly perceived as a security threat after the outbreak of the first Intifada (1987-1993). They have documented the way that movement restrictions (Hass, 2001) have reduced the number of work permits and marginalized Palestinian workers presence in the Israeli economy (Farsakh, 2005; Arnon *et al.*, 1997), a process that accelerated during the years 2000 to 2010 (Mansour, 2010; Etkes, 2011; Ekstein, 2011). Following the second Intifada (2000-2004) and the Gaza War (2006), the entry of Palestinians has been further limited by Israeli governments. In 2011, Palestinian workers represented only 2% (Ekstein, 2011) of the Israeli work force, compared with 7% in 1987 (Portugali, 1989).

Yet, in concentrating on the shrinking role of Palestinians within the Israeli economy, this literature has overlooked the parallel rise and professionalization of an informal economy of border crossing. As in

[1] I would like to thank Myriam Catusse (CNRS, IREMAM), Jean François Legrain (CNRS, GEMMO) and Sylvie Mazzella (CNRS, LAMES) for their reading and comments of this chapter.

many other border contexts (Andreas, 2001), the escalation of controls forced labourers to resort to new tactics in their attempts to transit the reinstated 'border zone' between Israel and the Occupied Territories. The first to focus on this development was the anthropologist Avram Bornstein who carried out fieldwork in the northern West Bank in the late 1990s. Already there were many more Israeli checkpoints and random patrols along the Green Line,[2] when Bornstein (2002) began his study of routes taken by Palestinian workers and contractors to reach work sites in Israel. Later research described how crossing into Israel became more difficult due to tighter controls (B'tselem, 2007a, 2007b). Palestinians improvised new tactics (Parizot, 2004; Kelly, 2006, Ch. 4), skills and knowledge in order to find a way through the cracks in Israel's surveillance system (Amiry, 2010; Bontemps, 2011). Others took advantage of business opportunities generated by the difficulties faced by Palestinian workers looking for jobs in Israel. In this way, they fostered the emergence of a new 'border economy' (Parizot, 2008).

The term 'border economy' refers first to the 'industry of crossing' (Hernandez-Leon, 2008) in which formal and informal entrepreneurs are *de facto* participants, alongside the State institutions, in the process of regulating mobility across boundaries. It also refers to new types of economic activities that emerge in and structure border areas. This is not simply a question of readjusting the shape and range of cross-boundary networks set up by workers, smugglers, employers, and state institutions: it also affects the nature of relationships linking these various actors by creating new power relations, antagonisms, and forms of exploitation (Parizot, 2004, 2008).

This article will pursue an earlier analysis of this 'border economy' in order to understand how it affects and contributes to readjustments to Israeli control mechanisms. Until now, there has been little interest in the implications of informal activities for Israeli control procedures. Scholars and observers in this field have mostly sought to document the hardship of daily life in Palestine and the way the Palestinian people resist and denounce injustices resulting from Israel's separation policy. My aim here is neither to neglect the suffering of Palestinians, nor to fall in with the discourse of resistance. On the one hand, observing human behaviour through the prism of resistance can often be misleading. John Comaroff (1993) has shown that using such an approach often leads to the observer's interpretation being projected onto the people observed.

---

2 The Green Line corresponds to the armistice line agreed by Israel, Egypt and Jordan at the end of the first Arab-Israeli conflict (1949). After Israel invaded the West Bank and Gaza, this line was recognized by the UN as separating Israel proper from the Palestinian Occupied Territories but has never been accepted by Israel as a legal border.

As Lila Abu Lughod (1990) states, it tends to overestimate the capacity of the powerless. It also reinforces a binary construct of the conflict, based on the opposition of Israeli against Palestinian, which neglects other significant interactions between these protagonists. On the other hand, by focusing solely on how Palestinian lives have been reshaped by Israeli controls, scholars limit their focus to the sociology of power.[3] By taking a different approach, i.e. by examining how Palestinians adapt and take advantage of faults and opportunities created by the Israeli surveillance system, we can obtain a more dynamic and nuanced understanding of the way this system operates.

My argument here is that, despite the small numbers of Palestinian workers smuggled daily, the emergence of this informal economy has deep ramifications for both Israeli and Palestinian societies. It establishes a variety of networks linking groups and actors that are usually thought to be independent. By taking into account the multiplicity of participants (Palestinians and Israelis, Jews and Arabs) in the border economy and the nature of their relationships with Israeli authorities, I shall demonstrate that they are in fact fully integrated in the mechanisms regulating the movement of Palestinians.

This chapter draws on ethnographic data I have collected since the mid-1990s in the southern part of the West Bank between the Jebel al-Khalîl (South Hebron Hills) and the northern Negev. Thanks to my ability to travel freely with a French passport and the networks I developed on 'both sides' of the Israel/Palestine divide, I was able to interview Palestinian workers, watchers, drivers, and contractors, as well as Israeli employers (Jews and Arabs living inside Israel and settlers in the West Bank), smugglers, and representatives of Israel's army and police. By comparing this data to other observations from previous visits to clandestine points of passage and taking into account prior research, I concluded that the dynamics observed in the southern West Bank are indicative of a wider trend found throughout the border zones between the West Bank and Israel.

I will first describe the changing nature of the borders between Israel and the Occupied Palestinian Territories and how this has affected the passage of Palestinian workers into Israel over the last 20 years. This will be followed by an assessment of how this new context has progressively favoured the emergence of a well-organized informal border economy. Finally, by focusing on the ramifications of this economy, I will examine how it has become an integral part of informal Israeli controls of Palestinian mobility.

3 I owe this idea to Alessandro Petti (Al-Quds University, Jerusalem).

1967-2011: from open borders to separation

Palestinians were first hired to work in Israel after the occupation of the West Bank and the Gaza Strip in 1967. Since then, their place in the Israeli economy and society has changed considerably, as have their conditions of living and border crossing. Between the early 1970s and 1987, they were fully integrated in the Israeli economy. The fact that most of them did not have work permits was not considered to be an issue (Portugali, 1989). However, after the first Intifada (1987-1993), they were seen as a growing security threat which led the Israeli authorities to impose tight restrictions as a way of regulating and monitoring their mobility. Though introduced as ad hoc security measures, these regulations became the basis for a systematic policy of separation.

From integration to separation

Contrary to Israeli's expectations, employment of Palestinian workers did not prevent them from rebelling against the colonial regime imposed after the Six Day War in 1967.[4] Between 1987 and 1993, Palestinian political factions called for a number of coordinated actions that would free them from economic dependence on Israel: general strikes, refusal to pay taxes, and boycotts of Israeli products. Palestinians were encouraged to stop working for Israeli employers in both the private and public sectors. These actions had a profound impact on both the Israeli and the Palestinian economy. In parallel with these campaigns of pacific resistance, there were confrontations between Palestinian demonstrators and the Israeli army in the West Bank and Gaza, including acts of sabotage and attacks against Israeli employers' worksites.

In order to counter this rebellion, the Israeli authorities resorted to different measures. Inside the Occupied Territories, they introduced curfews and closure (Hass, 2001). They also set up checkpoints along the main roads linking the West Bank and the Gaza Strip to Israel. Between 1989 and 1991, Palestinians working in Israel were required to carry magnetic cards and work permits (Hass, 2001; Abu Zahra, 2007). In 1993, the Israeli government created new quotas for Palestinians working in specific sectors (OECD, 2010) and entry into Jerusalem (which had been annexed in 1967) was restricted to permit holders.

Although these steps were originally introduced as ad hoc counterinsurgency measures, they gradually became an integral component in a more systematic policy to separate Palestinians from Israelis (Hass, 2002). The objective was not solely to prevent unwanted individuals

---

4 The Israeli authorities believed that maintenance of a certain level of prosperity and of Palestinian dependence on the Israeli economy would avoid such an outcome (Hass, 2001; Gordon, 2008; Grinberg, 2011; Bornstein, 2002).

entering Israel, but also to cut back on the number of Palestinian workers. It was thought that these steps would reduce Israel's dependency on Palestinian workers and thus the country's vulnerability to Palestinian resistance (Grinberg, 2011). By the late 1980s, Palestinians represented around 7% of the Israeli workforce (OECD, 2010, 213) and, in strategic sectors such as agriculture and construction, 49% and 45% respectively. It was not until massive numbers of workers arrived from Asia and Africa in the early 1990s, that the Israeli government could enforce restrictions on the movement of Palestinians (Kemp, 2004).

Two other factors favoured implementation of the separation policy introduced by the Labour government elected in 1992. First, in September 1993, the signature of the Agreement of Principle between Israel and the PLO presaged the gradual introduction of autonomy for Palestinians and, in the long term, the possibility of the creation of an independent Palestinian State (Grinberg, 2007; Latte Abdallah and Parizot, 2011). In this context, the international community began to recognize the legitimacy of steps already taken by Israel to end their responsibility for the economic well-being of the Palestinian population, particularly with the arrival of international aid and funding for the process of establishing a Palestinian State. Second, rising tensions between the Israeli government and the newly created Palestinian Authority, together with increased violence between the two populations[5] (Smith, 2007) gave the Israeli government additional reasons for reinforcing its policy of separation.

Yet, by the end of the 1990s, this policy did not in fact create a territorial division between Israeli and Palestinian territories. Rather, it fostered the emergence of a complex patchwork of Palestinian enclaves encircled by Israeli-controlled areas (Map 1). Even though Israel withdrew from Palestinian towns and delegated their administration to the newly created Palestinian Authority, it nevertheless reinforced its controls. Prior to the second Intifada (2000), the Palestinian Authority was responsible for security in only 18% of the West Bank. In B areas, which represent another 22%, it had to cooperate with Israel while the remaining 60% was fully under Israeli control.

---

[5] The first years following the signing of the Agreement of Principles in 1993 were particularly violent. On one side, Israeli settlers intensified their activities to take over Palestinian lands, in order to defeat the negotiations process. Some committed attacks directly on Palestinians, such as the Hebron massacre. Finally, in 1995, extreme right cells organized the assassination of Israel's Prime Minister Yitzhak Rabin. On the other, around 1994-1995, Palestinian factions began to carry out suicide bombings in the centre of Israeli cities.

Map M. Coulon, Lames, M. Barazani, CRFJ, C. Parizot, IREMAM. Mapping : ocha Information Managemant Unit, june 2009. Data base : Ocha, PA, MoP

Map 1: Location of A, B and C areas in the West Bank 2009)

Materialization of Separation and Control of Mobility

During the second Intifada (2000-2004), the Israeli authorities relied on a double-edged strategy: they reinforced the process of separation between the two populations and they attempted to regain security control over the Palestinian cities and villages from which they had withdrawn in the 1990s (Latte Abdallah and Parizot, 2011).

To achieve this objective, controls of the entry of Palestinians into Israel were further tightened. In fall 2000, the number of permits was reduced, which led a sharp drop of the number of Palestinian workers in Israel (see Table 1) which remained very low until 2004. Despite a slight increase since 2004, the number of permits issued in 2011 was still much lower than in 2000. In addition, the criteria of eligibility in terms of biosocial and security profiling have been redefined and became more stringent (Etkess, 2011). In 2002, the Israeli Prime Minister, Ariel Sharon, launched the construction of the Barrier/Wall[6] around the West Bank (Rabinowitz, 2003), as a way of preventing the passage of suicide bombers.[7] By creating a physical barrier, the number of 'points of passage' (*ma'avarim* [in Hebrew]) was reduced and security checks could be more thorough (Havkin, 2008, 2011). The construction of these 'terminals' moved the checkpoints further east, as many were placed along the Barrier, rather than on the Green Line (Havkin, 2008, 2011), as 85% of the route now lay inside the old armistice line (UNOCHA, 2011a). This also led to some controls being allocated to private companies which, since 2006, were now responsible for most of the checkpoints along the Wall (Havkin, 2011).

It was then decided to extend surveillance of Palestinian movements beyond the Wall by including all the C areas in the border zone. However, it was not possible to have uniform controls of movement in these areas, since different regulations were established for specific places. New military zones were identified and restrictions placed on Palestinians' ability to access Israeli settlements and bypass roads by Palestinians. In order to implement these restrictions, the Israeli army had to deploy additional road blocks near and around Palestinian enclaves. According to United Nations officers, the number of checkpoints, barriers, trenches, earth mounds, and road blocks remained relatively stable: 605 in February 2005 (end of the Second Intifada) and 589 in July 2010 (UNOCHA, 2005, 2010).

The southern West Bank offers a good illustration of the impact this policy had on the local population. From mid-1996 to 2002, there was only one checkpoint between the Negev and the Southern West Bank:

---

6 There are several terms for designating the Barrier that Israel has built around the West Bank. The choice of term depends on the speaker's political position: Israelis in favour of this project use the term 'security barrier' (*gader bitahon*) or 'separation barrier' (*gader hafrada*); Palestinians and their supporters call it the 'Apartheid Wall' or just the 'Wall'. Here, 'Barrier', 'Wall' and 'Fence' are indiscriminately as this construction is in fact a series of concrete walls, electronic fences, and patrol routes.

7 On the multiple functions of the Wall, see Latte Abdallah and Parizot (2011).

Meitar/Wadi al-Khalîl (Map 2), which monitored the passage of Palestinians and Israelis entering the West Bank from the south on Route 60. When the policy of confinement was launched in spring 2002, Palestinian villages and cities were temporarily re-occupied by the Israeli troops and long-term curfews and closures were introduced. In 2004, new checkpoints and roadblocks separated the South Hebron hills from Hebron and the rest of the West Bank. Later in the same year, after a suicide bombing in the Israeli city of Beer Sheva (August 2004), two additional checkpoints were installed south of the Samû'-Yatta enclave, blocking the two remaining roads giving access to Israel, and another checkpoint was added west of Ramadhin. In 2006, before completing the Separation Barrier along the Green Line, the army built an additional concrete wall (1.20 m high) along Route 317 to prevent access to Palestinian cars from the North. This 'obstacle' (*mekhshol* [in Hebrew]) was removed in 2008, but access to bypass roads remained restricted for private vehicles for at least another year. Even though, a period of relative calm in the region, between 2007 and 2010, led to many roadblocks being lifted near the Samû'-Yatta and Dûra-Dhahriyya enclaves, surveillance along and close to the Separation Barrier was reinforced.

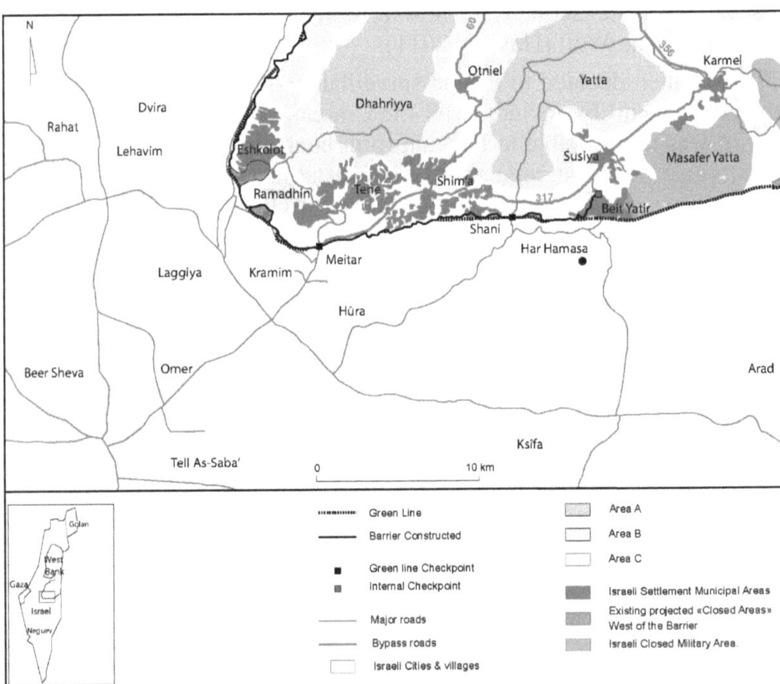

Map 2: Southern West Bank and Northern Negev (2010)

As in the rest of the West Bank, these measures had a dramatic impact on the mobility of Palestinians between local towns and villages and into Israel. In addition, the number of Palestinian workers employed in Israel in 2011 remained low compared to 2000: according to the Israeli Central Bureau of Statistics, the figure was 60,000, of whom more than 50% entered without legal papers (Ekstein, 2011, 10). Altogether, these workers represented 2.5% of the whole Israeli workforce and around 14% of workers in the West Bank. Even though the number of Palestinian employees in the Israeli job market has declined substantially since the first Intifada, they still hold a strategic role in some sectors such as construction and agriculture where they represent respectively 13% and 5.7% of the labour force.

Nevertheless, using statistics to assess the level of illegal crossings is problematic. First, these figures seem to underestimate the number of undocumented workers. Kav LaOved, an NGO advocating workers' rights in Israel, believes that they represent a much higher proportion of workers in Israel, with as many as twice the number of workers with permits (2009).[8] Second, institutional categories cannot be considered to be reliable analytical tools, especially since they aim to construct and dissociate groups of people according to their definition of what is legal or illegal status (De Genova, 2002). These workers do not live in a separate world: they are not merely interconnected, but they are often the same people. Palestinian workers with permits often stayed overnight in Israel in order to avoid the difficulties involved in crossing the Barrier. Thus, the same persons could be 'legal' workers during the day and 'illegal' aliens at night. Furthermore, many Palestinians purchased permits from Israeli go-betweens without actually having a place of work. They could show their permit at the checkpoint but in fact worked illegally in Israel. Many more people were involved in these informal practices than the statistics claim, particularly if one takes into account the large number of people involved in helping workers to circumvent Israeli surveillance.

The informal border economy and the division of labour

The development of a border economy began in the 1990s but, until the early 2000s, the techniques used to circumvent Israeli surveillance were somewhat makeshift (Bornstein, 2002; Kelly, 2006). When Israel tightened its surveillance during the second Intifada (2000-2004) and the post-Intifada period, the Palestinians were forced to rethink their meth-

8 URL: http://www.kavlaoved.org.il/media-view_eng.asp?id=2223 (Accessed on 23.2.2012).

ods of transport, reallocate tasks and discover new forms of coordination across borders.

From an open transport market to a highly fragmented sector

In the mid-1990s, a number of people became involved in helping Palestinians avoid checkpoints in the southern areas of the West Bank. For example, as few workers could afford to drive their own cars into Israel, Israeli Bedouins[9] and West Bank residents offered informal minibus services. Sometimes Jewish and Arab employers would collect their workers from their homes. As in the north of the West Bank, drivers and contractors would plan their journey according to the legal status of their passengers: if all the passengers had a permit, they could use the main roads, if not, they found ways to bypass the checkpoints (Kelly, 2006).

In the late 1990s, when Palestinian vehicles were not allowed to cross the Green Line, travel between Palestinian towns and Israeli work sites became more difficult. Previous systems broke down and new staging places had to be created. Most Palestinian workers had to choose whether to travel in an Israeli car, take the risk of entering in a Palestinian car (which was not allowed), or change to an Israeli vehicle before crossing the barrier. In the short term, these new restrictions allowed the Negev Bedouins to take control of the transport business in Israel and, to a certain extent, in the southern West Bank, since, as Israeli citizens, they could travel freely between Israel proper and C areas in the West Bank.

After 2000, the introduction of more obstacles forced drivers to limit their activities to specific sections and routes. First, Israeli Jewish contractors stopped driving into the Palestinian enclaves to fetch their employees, because they were worried by the increased violence, the army's reoccupation of the area, and a number of attacks against Israeli Jews. Those that were still willing to collect their undocumented workers preferred to pick them up only after they had passed the checkpoints into Israel.

Because they are also Palestinians, Israeli Bedouin contractors and drivers were not worried about attacks from West Bank Palestinians, but the Israeli Army's re-occupation of Palestinian areas in 2002-2003 made

9 Bedouins are descendants of the semi-nomad Palestinian population that remained in the Negev after the first Israeli-Arab conflict (1947-1949). They obtained the Israeli citizenship in the early 1950s. At the beginning of the 21st century, the majority was living in planned townships and the remainder inhabits unrecognized slums around Beer Sheva. Although cut off from their Palestinian kin and neighbours living in the southern West Bank and the Gaza Strip by the new political borders in 1949, they have maintained continuous social and economic relationships (Parizot, 2001, 2004).

it difficult for them to travel freely in the West Bank. Even after the army withdrew, their journeys were still hampered by road blocks, trenches, barriers and checkpoints. As a result, the Bedouin drivers also wanted to avoid picking up Palestinian workers until after they had crossed the Barrier.

Finally, it became increasingly difficult and more complex for Palestinian workers to travel to the checkpoints and be picked up by an Israeli driver. In 2007, a worker living in the periphery of Yatta could use his private car to go into the centre of town but once Route 317 was forbidden to private Palestinian cars, he had to take public transport to reach Dhahriyya. There he had to wait for a driver who would take him safely from Dhahriyya to the Fence by avoiding Israeli patrols. There an Israeli driver waited to take him to his destination. Thus in the late 2000s, he spent one to two hours to make the same journey that took only half-an-hour before the second Intifada, even without leaving the West Bank.

As a result, drivers transporting undocumented Palestinian workers began to limit their activities to their own neighbourhood. They operated in specific areas in the West Bank and, with so many people involved, this necessitated a much higher level of coordination.

Division of labour

Since 2006 and the partial completion of the Barrier south of Dhahriyya and Yatta, Israel was able to channel the passage of undocumented workers and concentrate on establishing control over a small number of illegal crossing points: the areas around Ramadhin/Eshkolot, just north of the Wadi al-Khalîl/Meitar border terminal and of Masafer Yatta close to the Dead Sea (see Map 2). By tightening patrols in these areas, the nature of illegal crossings changed radically: in the late 1990s and early 2000s, drivers could still *circumvent* Israeli surveillance by taking alternative routes, but by the late 2000s, they literally had to *confront* Israeli control, i.e., they either had to force their way past border guards and army barriers in order to reach the Barrier close to Ramadhin or cross the desert south of Yatta. In 2009-2010, large numbers of undocumented workers tried to cross into Israel on Saturday afternoons and evenings and this led to car chases between Israeli jeeps and Palestinian drivers (*shoferiyya* [in Arabic]). Drivers had to improve their driving skills if they wanted to outrun Israeli jeeps on a very rough terrain. As a result, crossing the barrier became increasingly dangerous with more Palestinian workers and their drivers being wounded or killed on these routes than in the past.[10]

---

10 Forcing their way through Israeli patrols is not a specificity of the southern West Bank Palestinians. Suad Amiry (2010) has provided rich descriptions of this phe-

With a few exceptions, these drivers were no longer workers, contractors, or even workers engaged in smuggling. In the Yatta region, most were unmarried male adolescents or young men who wanted to take up this lucrative, but dangerous, business. Most came from a few large families already involved in this business, but some were recruited from other groups. On the Israeli side, the Bedouins also appeared to recruit additional drivers but, given risks involved, there was often a lack of volunteers.

Drivers needed immediate and continuous updates on the whereabouts of the Israeli army, police and border guards. They had to set up networks of watchers or scouts who were sent ahead of convoys to check the roads for patrols. These were local residents or workers employed by smugglers. Payment was sometimes in kind: for example, Sâlim Abu Jaffal[11] who worked in the Ramadhin/Eshkolot region in 2010, received prepaid SIM cards as payment for calling drivers from his lookout.

Smugglers also relied on help from local peasants and their excavating equipment. They regularly needed to maintain and open up new routes to avoid Israeli surveillance, and to remove earth-mounds or refill trenches made by the army to block roads. As a result, these activities reshaped the local landscape by regularly readjusting the network of dirt roads along the Barrier.

Last but not least, smugglers made use of local shopkeepers in towns along the routes taken by undocumented workers. In Dhahriyya, restaurants and hotels provided food, accommodation, and staging places where workers would gather before crossing. Local garages provided services and storage for unlicensed cars used for travel between the town and the Barrier. In addition, similar forms of cooperation were established inside Israel proper.

Business cooperation across borders

Smugglers needed tight coordination for their activities across the separation line between Israel and the Occupied Territories. Nevertheless, allocation of territory for their activities did not lead to a fragmentation of the border economy, but rather resulted in a high level of rational and calculated integration.

---

nomenon in the region of Qalqilya where Palestinian workers cross by foot. The Israeli NGO, B'tselem, has posted a series of videos of these daily events. See e.g. *Catch me*, 2007, URL: http://www.btselem.org/video-channel/movement (Accessed 29.5.2012).

11 All names in this article have been changed to protect the privacy of interviewees.

In 2010, in the area between the Southern West Bank and the North-East Negev, Palestinians and Bedouins maintained constant contact in order to agree on specific times and places where they could pick up and drop off workers along the Fence. Moreover, Bedouins needed to coordinate with other transporters (such as Israeli Arab or Jewish taxis and minibuses) who would then take Palestinian workers from local Bedouin towns to work in southern, western and northern Israel.

In some regions, for example Dhahriyya/Ramadhin, smugglers on both sides worked together to force their way through Israeli patrols. In summer 2010, Palestinian drivers would lie in wait with their Mitsubishi truckloads of workers on the hill dominating the Barrier. Once informed by watchers that Israeli jeeps blocking the road had moved, they would rush downhill as fast as possible towards a hole in the Fence. On the other side, Bedouin drivers would drive down to the Barrier and pick up the workers. By the time the Israeli jeeps hurried back, both sets of drivers had already disappeared.

Coordination of drivers involved not only well-timed simultaneous operations but also the right number of cars on each side: Palestinian trucks could carry up to a dozen workers, but Bedouins drive cars that take only 5 to 7 people, doubling the number of cars required.

These manoeuvres also involved coordinated financial transactions since workers only paid drivers when they reached their destination. A worker traveling from Dhahriyya to a Bedouin township in the Negev paid a fee (NIS 150)[12] to the Bedouin driver. Conversely, a Palestinian returning to Dhahriyya paid the Palestinian driver. These earnings had to be redistributed among the drivers on both sides of the border.

Overall, these groups function as 'entrepreneurs without borders', not just because they are involved in business cooperation across the barrier, but also because their activities and practices are regulated by informal contracts which apply to local groups operating on both sides in order to avoid sources of conflict. Coordination of their activities was particularly important, because the very lucrative nature of their business has attracted many newcomers and generated fierce competition among smugglers. In 2010, local Bedouin elders in Israel decided to force the various extended families involved to come to an agreement. The objective was twofold: fix the rules for people entering this lucrative market, and punish certain recurrent behaviour, including denunciation of competitors to the Israeli police.

---

12 In 2010, approximately €30, the equivalent of one day's salary in Israel for an unskilled Palestinian construction worker.

Ramifications of the informal border economy

Over the last ten years, despite the small number of workers involved, informal crossings have not merely generated well-integrated business activities and partnerships, they have also had profound ramifications for the local business and social environment on both sides of the barrier, and in communities living near the Fence and further inside the West Bank. The impact of these ramifications is all the more impressive when we take into account the number of people who have tried to capitalize on these networks and crossing activities for their own purposes.

Smugglers' partners

Border entrepreneurs are very dependent on their contacts and acquaintances for obtaining essential information and this generates widespread and varied information networks involving both Palestinians and Israelis. In addition to watchers and lookouts along West Bank roads and at checkpoints, smugglers also rely on their contacts within the Israeli administration.

Bedouin drivers involved in smuggling Palestinian workers often belong to extended families which have Israeli police officers, border guards and soldiers among their relatives. As a result, they have opportunities for obtaining useful information on the movement of Israeli patrols and the location of random checkpoints and allows them to adjust their plans accordingly.

Israeli settlers in the West Bank can sometimes play an important role as informers. They have greater ability to travel freely and this gives them very precise information on routes, obstacles and patrol movements. They often have developed good contacts in the army, because they have soldiers and officers defending the settlements billeted in their homes or, since the mid-1980s, they are sometimes recruited as 'territorial defence auxiliaries' (Gordon, 2008). In order to keep their precious cheap manpower, those who employ Palestinian workers illegally do not hesitate to warn them and their drivers about the whereabouts of Israeli patrols. Furthermore, some take advantage of their privileges to smuggle Palestinians through Israeli check points. In autumn 2008, a 63 year-old settler from South Hebron Hill was arrested at a check point with two Palestinians hidden in the trunk of his car.

After crossing into Israel, these networks are critical. They help to give undocumented workers some protection by providing them accommodation and hiding places. Their assistance is also crucial for gaining access to worksites; in 2008, a security guard responsible for employment procedures near Jerusalem received regular bribes from contractors to allow the entry of workers without proper permits (Glickman, 2008). They also prevent arrests of undocumented teams on work sites. Between 2007 and 2010, thanks to his contacts, Ahmad Ga'ûd, a Palestinian contractor, was able to hold on to his team of undocumented Palestinian workers in a major meat factory within a Negev industrial area. Because the mother of his Israeli partner was a local policewoman and because the owner of the factory had contacts in the Israeli administration, Ahmed Ga'ûd was confident that there would be few police patrols on his worksite. In 2008, he even managed to keep his team working on Yom Kippur when everybody stops working in Israel!

Retribution, extortion and the informal tax system

Exchanging information and good practices comes at a cost, whether social or financial. Bilâl Abu Kammûne, a 30 year old Bedouin contractor from Rahat, became friendly with Israeli soldiers stationed at the Meitar/Wadi al-Khalil checkpoint before it was handed over to a private company, by inviting them several times to bars and local restaurants in Beer Sheva. He managed to win their confidence while convincing them that he needed to employ undocumented Palestinian workers. As a result, he received a pass for the Meitar check point. In return for their generosity, he delivered cigarette cartons. After 2006, when the army and police officers were gradually replaced by private security guards on most West Bank checkpoints (Havkin, 2011), these forms of corruption continued: for example, in 2008, with help from local private guards, several merchants from Hebron were able to smuggle their goods through southern checkpoints.

In some instances, Israeli civil servants extort kickbacks from Palestinian workers. Taking advantage of their acquaintances among Palestinian workers and their ability to provide passes, they obtain useful services for free or a very low fee. In 2005, workers from Samu' were summoned by a Bedouin border guard living in Beer Sheva to repaint his house. He went to fetch them from the West Bank and took them across the checkpoint in his police car. After they finished, he asked them to leave immediately without payment and threatened to arrest them for entering Israel illegally!

Israeli soldiers, civil servants and private guards are not the only people to capitalize on their power and position. In 2009, a Bethlehem worker described how West Bank Palestinians in contact with smugglers were regularly paid for providing critical information or making contact with undocumented workers. Some even exploited Palestinian workers. In 2008, Palestinians living in Qatanna (north-east of Jerusalem) and working in the Israeli village of Nattaf were regularly forced to provide information on the whereabouts of their employers to a group of Palestinian thieves so that they could plan and time their robberies during their absence.

These forms of extortion are neither exceptional, nor arbitrary; they are part of a system created by an informal border economy. As in many other border contexts, such payments are in fact an informal tax for passage (Andreas, 2001): it is now almost obligatory and workers must be prepared to see this as part of the cost of crossing the border. This is a clear sign that the market is both dynamic and lucrative and that it attracts a multitude of actors, including State authorities. By assisting Palestinian workers and contractors to cross into Israel and extorting taxes and/or services from them, a whole range of 'regulatory authorities' – smugglers, private guards, Israeli civil servants and others individuals – have become part of this power mechanism (*dispositif de pouvoir*). Depending on their personal interests, they are in a position to open and close access to Israel and its job market.

Informal regulatory authorities under State control

The emergence of these informal controls has not weakened Israel's control of borders; rather, we can see that it has the capacity to redeploy its controls via informal systems.

Contrary to official political positions with regard to Israeli and international public opinion, which emphasize the mobilization of State authorities in the fight against Palestinian 'infiltrators' (*mistanenim* [in Hebrew]), Israeli civil servants have shown a certain degree of tolerance for the crossing of undocumented Palestinian workers, not least because they know that they do not have sufficient resources or manpower to stop smugglers. In 2008, in the region north of Jerusalem, an Israeli border patrol officer stated that 1,500 Palestinians, on average, jumped the Fence each week, but that he and his colleagues caught no more than 10% of them, due to their lack of resources. Moreover, because the lack of cells and staff at local police stations, border guards could not afford to charge the Palestinian workers they arrested. In addition, Israel does not have enough magistrates to judge or enough jails to incarcerate the Palestinian workers crossing without a permit (B'tselem, 2007b). Finally, Israeli authorities know that they cannot arrest all the smugglers. The business is so lucrative that dismantled networks will immediately be replaced by new groups.

For a long time, the General Security Services (*Shabak*), the police and the army have opted for infiltration of smuggling networks because they provide crucial information on Palestinian activities. This has high priority as a way to prevent suicide bombings and weapons trafficking. Smugglers already have good connections with police officers, soldiers and private security guards whom they attempt to corrupt in order to obtain crucial information for avoiding Israeli surveillance. In addition, the intelligence services often force these officers to play a double game while, at the same time, they recruit Palestinian workers, contractors and smugglers (B'tselem, 2007b). With the large number of interconnections in such a vast network, a few key well-positioned collaborators can obtain information on the arrival of unidentified participants. If arrested, undocumented workers and smugglers can face several years of imprisonment and very high fines. Police, *shabak* and army officers are often in a position to force smugglers and workers to collaborate in return for their freedom.

Instrumentation of smuggling networks gives the Israeli authorities more resources for curbing suicide attacks inside Israel as it allows them to monitor the networks used by Palestinian factions during the second Intifada. Besides, this system of informal controls has proved to be very efficient. Because they know they are under surveillance, smugglers are very careful in choosing their clients. Besides they often require newcomers to submit to intrusive interviews and thorough checks in order to be sure that they are not suicide bombers. Finally, by tolerating and controlling crossings by undocumented Palestinian workers and the informal economy that this has generated, Israeli authorities have managed to maintain and reinforce their control over border zones, Palestinian enclaves and Israeli territory.

Conclusion

Since the 1990s, the escalation of security measures and movement restrictions in the border zones separating Israel from the Occupied West Bank has reduced, but not stopped, the passage of Palestinian workers entering Israel. It has merely changed the way they cross the divide, not only by making the passage harder, but also by fostering the emergence of an informal border crossing economy.

This economy has deep ramifications because of its impact on the local population, over and beyond workers and smugglers involved. There are now highly inventive entrepreneurs who have taken advantage of the opportunities created by the system of constraints imposed by Israel over the last 20 years. Their business relies on collaboration with a variety of local actors: watchers, drivers, building labourers, car mechanics, restaurant owners, etc. Outside the border zones, the sector functions thanks to the complicity of certain Israeli civil servants and private entrepreneurs. For an informal market, it is particularly well-integrated and regulated with a pragmatic and calculated distribution of spheres of influence and division of labour that requires a high level of coordination across borders, both in circumventing Israeli systems of surveillance and in managing financial issues. Finally, the participants have negotiated a structured framework governing competition and entry into the market, with assistance from recognized local authorities.

Over this period, this informal economy has evolved into a mechanism of movement regulation alongside governmental structures. Not only does it facilitate the movement of undocumented workers, but it also fixes prices and collects informal taxes. Furthermore, it has led to a reorganization of roads and staging places on the routes taken by Palestinians travelling between home and work. However, the State continues to maintain a certain level of control over the system: many smuggling networks are known and utilized by Israeli intelligence services and play a key role in the mechanisms for regulating Palestinian mobility. We must therefore reconsider how the system of border controls imposed by Israel on West Bank Palestinians operates as a heterogeneous power mechanism (*dispositif de pouvoir*) covering public and private, and formal and informal, actors whose activities are not necessarily coordinated, but highly synchronized.

References

Abu Lughod Lila. 1990. "The Romance of Resistance". *American Ethnologist* 17(1): 41-55.

Abu Zahra Nadia. 2007. *Legal Geographies in Palestine: Identity Documentation, Dispossession, Repression and Resistance*. PhD thesis, Geography, University of Oxford, under the supervision of Tony Lemon and Dawn Chatty.

Amiry Su'ad. 2010. *Nothing to Lose but your Life*. Doha: Bloomsbury Qatar Foundation.

Andreas Peter 2001. *Border Games: Policing the US-Mexico Divide*, Ithaca: Cornell University Press.

Arnon Arie, Israel Luski and Jimmy Weinblatt. 1997. *The Palestinian Economy. Between Imposed Integration and Voluntary Separation*. London: Brill.

B'Tselem. 2004. *Forbidden Roads: The Discriminatory West Bank Roads Regime*. Jerusalem.

B'Tselem. 2007a, *Ground to a Halt: Denial of Palestinians' Freedom of Movement in the West Bank*. Jerusalem.

B'Tselem. 2007b, *Crossing the Line: Violation of the Rights of Palestinians in Israel without a Permit*. Jerusalem.

B'Tselem, 2007c. *Catch me*, [online video] URL: http://www.btselem.org/video-channel/movement (Accessed 29.5.2012).

Bontemps Véronique. 2011. "Narratives of Palestinian workers' employed in Israel", in A. Khalil (ed.) *Palestinian Refugees: Rights, Narratives and Policies*. Birzeit University: 219-232 (in Arabic).

Bornstein Avraham. 2002. *Crossing the Green Line between the West Bank and Israel*, Philadelphia: University of Pennsylvania Press.

Comaroff John and Jean Comaroff. 1993. *Resistance and Rebellion in Black South Africa, 1830-1920: Awful Acts, Lawful Facts, and Illegal Imaginings on a Colonial Frontier*. Research Proposal funded by the NSF and the American Bar Foundation.

De Genova Nicholas P. 2002. "Migrant 'illegality' and deportability in everyday life". *Annual Review of Anthropology*, 31: 419-447.

Ekstein Zvi. 2011. *Report of the Committee for Regulating and Monitoring the Employment of Palestinians in Israel*. May 2011. Jerusalem, Cabinet Secretary's Office (in Hebrew).

Etkes Haggay. 2011. *The Impact of Employment in Israel on the Palestinian Labor Force*. Discussion papers No. 2011.11. Jerusalem, Research Department, Bank of Israel.

Farsakh Leila. 2005. *Palestinian Labour Migration to Israel: Labour, land and occupation*. London: Routledge.

Glickman Aviad. 2008. "Guard who allowed Palestinians into Israeli town without permit sentenced", *Ynetnews.com*, October 30, 2008 [online article] URL: http://www.ynetnews.com/articles/0,7340,L-3615403,00.html (Accessed 31.10.2008).

Gordon Neve. 2008. *Israel's Occupation*. Berkeley: University of California Press.

Gringberg Lev Luis. 2007. *Imagined Peace*. Ram Allah, Mu'assasat al-ayâm, Madar (in Arabic).

Grinberg Lev Luis. 2011. "Discours économiques et construction des frontières dans l'espace israélo-palestinien depuis 1967", in S. Latte Abdallah and C. Parizot (eds.), *A l'ombre du Mur: Israéliens et Palestiniens entre séparation et occupation*. Arles: Actes Sud: 105-123.

Hass Amira. 2001. *Boire la mer à Gaza*. Paris: La Fabrique.

Hass Amira. 2002. "Israel's closure policy: an ineffective strategy of containment and repression". *Journal of Palestine Studies*, 31(3): 5-20.

Havkin Shira. 2008. *La privatisation des checkpoints: esquisse d'une transformation d'un dispositif de pouvoir*. Master 2 Thesis in Philosophy, supervised by Jocelyne Duboist and Elsa Dorlin, Université Paris 1.

Havkin Shira. 2011. "La privatisation des check-points: quand l'armée rencontre le néo-libéralisme", in S. Latte Abdallah and C. Parizot (eds.), *À l'ombre du Mur: Israéliens et Palestiniens entre séparation et occupation*. Arles: Actes Sud.

Hernandez-Leon Ruben. 2008. *Metropolitan Migrants*. Berkeley: UCP.

Kelly Tobias. 2006. *Law, Violence and Sovereignty among West Bank Palestinians*. Cambridge: CUP.

Kemp Adriana. 2004. "Labor migration and racialisation: Labor market mechanisms and labor migration control policies in Israel". *Social Identities*, 10 (2): 267-292.

Kemp Adriana and Rebeca Raijman. 2008. *Migrants Workers: The Political Economy of Labor Migration in Israel*. Tel Aviv: Hakibbutz Hameuchad (in Hebrew).

Latte Abdallah Stéphanie and Cédric Parizot. 2011. "À l'ombre du Mur: comprendre le régime d'occupation israélien", in S. Latte Abdallah and C. Parizot (eds.), *À l'ombre du Mur: Israéliens et Palestiniens entre séparation et occupation*, Arles: Actes Sud: 15-48.

Mansour Hani. 2010. "The effects of labor supply shocks on labor market outcomes: evidence from the Israeli-Palestinian conflict". *Discussion Papers* 1014. Berlin, German Institute for Economic Research. Online: http://ssrn.com/abstract=1639828.

OECD. 2010. *Israel, OECD Reviews of Labour Market Series*. Secretary-General of the OECD.

Parizot Cédric. 2001. "*Gaza, Beersheba, Dhahriyya*: another approach to the Negev Bedouins in the Israeli-Palestinian space". *Bulletin du Centre de recherche français à Jérusalem* [Online], 9 | 2001, Uploaded on 25.2.2008, Accessed August 29, 2012. URL: http://bcrfj.revues.org/1742.

Parizot Cédric. 2004. "Crossing and Constructing Borders within Daily Contacts". *Notes de Recherche du CER*, No. 287-(2004/10).

Parizot Cédric. 2008. "Tightening closure securing disorder". *Refugee Watch*, 31, 54-74.

Portugali Juval. 1989. "Nomad labor: theory and practice in the Israeli-Palestinian case". *Transactions of the Institute of British Geographers*, 14(2): 207-220.

Rabinowitz Dan. 2003. "Borders and their Discontents Israel's Green Line, Arabness and unilateral separation". *European Studies*, No. 19.

Smith Charles D. 2007. *Palestine and the Arab-Israeli Conflict.* Boston, Bedford/St Martin's.

UNOCHA. 2005. *Humanitarian Update, Occupied Palestinian Territories*. Online: http://www.ochaopt.org/documents/OCHAoPt_ClsrMp_WB_En0805.pdf.

UNOCHA. 2010, *West Bank and Gaza Strip Closure Maps*. Jerusalem: UNOCHA.

UNOCHA. 2011a. *Palestinian Communities affected by the Barrier, Map.* Jerusalem: UNOCHA.

Border Economies

A Nascent Migration Industry Around Lampedusa

Heidrun FRIESE

Situated between Tunisia and Sicily, the tiny island of Lampedusa has become a highly visible symbol for Mediterranean borderlands, EU migration policies and cross-border governance.[1] Although the right to mobility is included in the Universal Declaration of Human Rights, undocumented people encounter – not least following the introduction of the Schengen Treaty and a visa system in 1986 – legal systems and border regimes that limit their freedom of movement and capacity to stay. Under pressure from the EU, several countries have signed bilateral accords which have led to the crossing of borders without legal documents becoming a legal offence: Morocco in 2003, Algeria in 2008 and Tunisia in 2004 (Ben Cheïkh and Chekir, 2008: 6-8; Fargues, 2009). However, externalization of European borders to former colonies has not stopped undocumented migration. Since the Tunisian Revolution for Freedom and Dignity (*Thawrat al hurriyya wa al-karâma*) and the fall of the Gaddafi regime in Libya in 2011, around 60,000 people have protested against this refusal of their right to mobility and crossed the Mediterranean to land at Lampedusa.

A multitude of local and supranational actors are involved in undocumented migration and the border regime. Their practices are similar but not dictated by any central logic or rationality. The notion 'regime':

> allows rooms for gaps, ambiguities and outright strains: the life of a regime is a result of continuous repair work through practices. Finally, the idea of a "migration regime" helps to stress the interdependence of observation and action (Sciortino, 2004:32).

The border regime also brings together a series of shifting relations between the global and the local and between the centre and the periphery and sets up incessant processes and shifting textures of de/re-

[1] The following remarks are part of a multi-sited research project on *The Limits of Hospitality* focusing on undocumented mobilities (Lampedusa, Sicily and Tunisia). The *Deutsche Forschungsgemeinschaft* has generously funded the project. For a more detailed account of the historical semantics of hospitality and its tensions, see Friese 2004, 2010, 2011.

localizations, de-bordering and re-bordering which blur the line between internal and external borders (Mezzadra, 2006), and between formal and informal economies. The border regime creates links among various actors and networks: on the one hand, those who burn their papers (known as *harraga*), their friends, families and networks, together with self-made entrepreneurs who arrange their voyages; and on the other, those who are required to manage their reception, such as members of the coastguard and security forces, employees of reception centres, local municipal council and other political actors. The border regime also leads to a growing presence of competing NGOs (e.g. International Catholic Migration Committee (ICMC), International Organization for Migration (IOM), International Centre for Migration Policy Development (ICMPD), Save the Children, Médecins sans Frontières (MSF)), public-private enterprises, research institutes, and other organizations. As a result, there is a dense international collection of actors, practices, policies, and powerful images of undocumented migration. References to 'flow', 'invasion', 'crisis', 'invasion' or 'emergency' are omnipresent and play an important role in this industry, while the media seek out and disseminate illustrations of the dominant attitudes towards undocumented migration. In these media campaigns, migrants are simultaneously victims and threats to national security and welfare systems. As a result, they have been absorbed into a powerful social *imaginaire* as 'a catastrophe' and this further legitimizes the political discourse and reactions. The emergency is no longer the exception, it has become the rule (Agamben, 1995).

Shifts in the concept of sovereignty and in the multiplicity of power relations, technologies, programmes and strategies involved in liberal and neo-liberal governance have been critically analyzed (Barry *et al.*, 1996). Research has also acknowledged the introduction of new types of hospitality (Derrida and Dufourmantelle, 1997), the quest for cosmopolitan concepts (Honig, 2001; Benhabib, 2004) and the need of new forms of citizenship (Bauböck, 1994; Carens, 1995). However, so far, economic issues relating to border regimes – specifically political economies and the nascent migration industry, relations between local and translocal/transnational informal and formal economies – have largely escaped attention. The 'production of illegality' (de Genova, 2002) has not only contributed to the brutal exploitation of illegal workers, it has also allowed security services to evolve into an emergent market which provides technically sophisticated devices for the detection of people crossing borders without documents (see Walters, 2006: 169-67).

At the same time, border regimes foster the production of subjectivities, the emergence of cross-border actors and the development of strategies to circumvent – or actively take advantage of – established procedures and routines for individual migratory journeys. Borders have become not only a means of exclusion and division, but also porous zones where contacts, encounters, border commuters and in/formal commerce (whether legal or illegal) provide a livelihood for inhabitants of these regions.

In this article, I will first review the various agents who are developing (new) forms of governance for undocumented migration and the massive economic interests that are at stake for supra-national and local agents (section 1). Against this background and based on a multi-sited anthropological research on Lampedusa, Sicily and ports in Tunisia, I will describe the intersections between formal and informal economies. As the political economy of borders (section 2) is closely related to powerful visualizations of catastrophes allegedly caused by migrants, its role in the creation of a permanent state of emergency will briefly be discussed (section 3).

Governing undocumented mobility and economical interests

Issues relating to (undocumented) mobility are now a core element in political and economic relations between the EU and its neighbours in the Mediterranean Basin. Currently, European policy promotes a two-fold approach to the containment of movement and undocumented mobility: the externalization of national borders and tighter controls and their multiplication within the national territory. Borders, as William Walters (2006: 197) has observed, "operate like filters or gateways", selecting "the good and the bad, the useful and the dangerous, the licit and the illicit; (...) immobilizing and removing the risky elements so as to speed the circulation of the rest."

Thus, cooperation and actions for the externalization of mobility controls in the Mediterranean and within former colonies have been upgraded. There is now a dense fabric of think tanks, research institutes and NGOs seeking to influence the political agenda and competing for donors and public funding. The most visible organizations operating in the Mediterranean are: International Catholic Migration Committee (ICMC), International Organization for Migration (IOM) and International Centre for Migration Policy Development (ICMPD). In addition, when the '5+5 Dialogue on Migration in the Western Mediterranean' was established in 2002, a number of organizations were accepted as 'partner and observer institutions' – IOM, ICMPD, and International Labour Organization (ILO) – in order to put together a political forum for "information exchange, joint management of international borders, agreed forms of labour migration, migration for development, and protection of the rights of migrants."[2]

Maintaining "partnerships with relevant governmental, multilateral and private sector industry partners",[3] IOM was one of the initiators of externalized borders and it implements the EU policies that it has sought to influence. Its decentralized organizational structure does not have an annual operating budget, but is financed via project-related funding and donations from its member states. It is one of the major organizations providing service and migration management in an expanding sector and its "membership increased from 67 States in 1998 to 146 States in 2012 and continues to grow. Total Expenditure increased from US$242.2 million in 1998 to US$1.3 billion in 2011."[4] Given its structure, it "depends upon the funds it manages or generates" (Caillault, 2012: 143). During the Arab Revolutions for example, EU and IOM extended their "cooperation to stabilize at-risk communities and enhance migration management to support transitions in Egypt, Tunisia and Libya" (IOM, 2011). In December 2011, they signed a contribution agreement in Tunis, "whereby the European Union (EU) entrusted IOM with €9.9 million" for a three year programme to "support transition processes and promote sustainable recoveries in Egypt, Tunisia and Libya. More specifically [...] IOM will seek to address emerging migration-related challenges in each target country" (*Idem*).

By this time, multilateral cooperation with Libya and Tunisia was now an established tradition since Tunisia hosted the first meeting of the 5+5 dialogue in 2002 and Libya became a member of IOM in June 2004.

> Since 2003 the IOM and Libya have participated in a series of technical and policy consultations, resulting in the development of a one-year plan of action, scheduled to begin in late 2004, under the Programme for the Enhancement of Transit and Irregular Migration Management in Libya – TRIM (Hamood, 2006: 24).

TRIM envisaged voluntary programmes for "irregular migrants" and the "enhancement of three reception centers" (Hamood, 2006: 24) which

---

2 http://www.iom.int/jahia/Jahia/pid/860 (Accessed 7.6.2012). 5+5 members are Algeria, France, Italy, Libya, Malta, Mauretania, Morocco, Portugal, Spain and Tunisia. Tunisia chaired the first meeting in 2002, the following meetings were held in Rabat (2003) and Algiers (2004).

3 IOM, http://www.iom.int/jahia/Jahia/about-iom/organizational-structure/migration-management-services (Accessed 4.10.2012).

4 IOM, http://www.iom.int/jahia/Jahia/about-iom/organizational-structure/lang/en (Accessed 4.10.2012).

would promote the exterritorialisation of the detention system to the former colonies.

In 2002, ICMPD encouraged the intergovernmental agency, Mediterranean Transit Migration Dialogue (MTM Dialogue), to intercede between "migration officials in countries of origin, transit and destination along the migration routes in Africa, Europe and the Middle East" with regard to irregular migration. In 2009, the MTM Dialogue entered the fourth phase of its activities, "a dialogue in action", following the guidelines "of the main global, African and European policy' in combating trafficking and smuggling".[5] MTM collaborates with the Global Forum on Migration and Development (GFMD), the Union for the Mediterranean (EuroMed), the Africa-EU Migration, Mobility and Employment Partnership (MME), the Rabat Process and the 5+5 Dialogue. Its guiding principles are explicitly designed to "contribute to the opinion-forming of state officials" and to operate "at the technical level."[6]

Border management systems and technologies of '*gouvernementalité*' (Foucault, 1994) are designed to detect and contrast resourceful strategies of border-crossings (even before they occur) and to detain mobile people. The production of knowledge is therefore a remunerative resource and a good product in the highly competitive market. "In order to elaborate knowledge-based approaches to migration, be this at policy level, strategic level, or through concrete actions, *access to comprehensive and updated information* is a prerequisite [emphasis in the original]" declared ICMPD.[7] There is a long tradition of using statistics, screening, mapping and visualization in developing techniques of surveillance, policing and *gouvernementalité*. For example, ICMPD's web site offers an interactive map – known as *i-map* – which provides visualizations of movements, routes, hubs and flows.[8] ICMPD's partners in implementing these maps on irregular and mixed migration are Europol, Frontex, Interpol, UNHCR and UNODC, Migration and Development: IFAD, IOM.[9]

Bilateral agreements between countries are a key element in the border regime and in efforts to control mobility. In August 2004, relations between Italy and Libya improved and the EU embargo of weapons was

5 ICMPD, http://www.icmpd.org/MTM.1558.0.html (Accessed 4.10.2012).

6 *Idem*.

7 http://www.imap-migration.org/index.php?id=4 (Accessed 4.10.2012).

8 For a broader view, see Tsianos (n.d.).

9 Its main donors are the European Commission and co-funding states, such as France, Italy, Malta, the Netherlands, Norway, Poland, Switzerland, UK (https://www.imap-migration.org/ (accessed 4.10.2012).

lifted in October, allowing Libya to buy modern surveillance systems including speedboats and to train its police forces. As a result of these accords, more than 1,000 people were deported from Lampedusa to Tripoli (an action highly criticized by UNHCR, Human Rights Watch and Amnesty International in 2005). In December 2007 another agreement was signed by Italy and Libya, which provided for joint collaboration in several areas including the fight against "criminal organizations involved in human trafficking and exploitation of clandestine immigration" (Ministero dell'Interno, 2007). Given the success achieved with the Italian-Albanian cooperation, collaborative patrols were expected to shut down these routes.

The EU border agency Frontex has become one of the key players in migration management. The agency is 'intelligence-driven' and provides additional 'border management systems' to Member States.[10] As the European Commission (EC) is Frontex's "most important donor at present, funding more than 95% of the agency's budget" (Kasparek and Wagner, 2012: 175): since "its inception in 2005, its annual budget has increased rapidly from €6 million to €82 million today. The Agency employs 290 staff" (Frontex, 2011: 4).

Frontex is mentioned in the Italian-Libyan protocol (a 'treaty of friendship' between Italy and the Gaddafi regime). Its representatives have carried out a mission in Libya and visited several detention camps for irregular migrants. These camps were found to be extremely poor yet, somewhat ironically, "throughout the mission, the members of the mission were treated to extremely warm hospitality by the Libyan hosts and useful operational contacts have been established which it is hoped will advance future cooperation on border security issues" (Frontex, 2007: 3). After the downfall of the regime – which was not known for respecting human rights – bilateral contacts between Italy and the provisional government have been re-established[11] along similar lines to contacts with Tunisia, not least with a view on containing mobility.

---

10 http://www.frontex.europa.eu/ (Accessed 2.11.2008).

11 The protocol of a recent meeting (3 April 2012) between the Ministers of the Interior of Italy, Annamaria Cancelleri and of Libya, Fawzi Altaher Abdulali, further affirms bilateral efforts to counter 'illegal' migration. It refers to the Palermo Convention against organized crime (2000) and the bilateral accords against 'human trafficking'. Its aim is to strengthen police collaboration, 'cross-border information systems' and establish a First Aid Centre for illegal immigrants at Kufra and other centres in Libya (Ministero dell'Interno Italiano, 3.4.2012). Following the European Court for Human Rights's judgment against Italy in 2012 however, collaboration between the two countries with regard to *refoulement* was – explicitly at least – not included in the protocol of mutual understanding.

These few examples illustrate a multi-level system of governance, the growing role of non-governmental actors who compete with each other and have a fundamental economic interest in the 'management' of mobility and implementation of migration policies. At the same time, they generate 'problems' and offer 'solutions' to the problems they placed on the political agenda in the first place. According to this remunerative logic, fighting undocumented mobility means fighting human trafficking and, subsequently, financing programs against trafficking becomes legitimate, even highly acceptable from a moral perspective and very welcome from a political perspective. Scientific discourse, political agenda-setting and fundraising in a competitive "market-segment" are all part of the dynamics of the border regime and conceptions of mobility. In the same vein already:

> the concept of transit migration was only invented during the 1990s and publicized by certain institutions, notably IOM, ICMPD, the Council of Europe and various UN agencies (Düvell, 2008).

As part of specific border regimes, 'migration management' comprises a set of discourses, images and practices carried by a multitude of actors who become intertwined in a highly complex, heterogeneous, competitive and often antagonistic, conflictive manner. It is also a highly profitable business at the intersection of formal and informal economies. At the same time, these practices, discourses and images relate to local actors, economical interests and conflicts about resources.

The political economy of borders

Harga, as a process, and the border regime are made up of specific social spaces and (in)formal economies. Despite the dominant political discourse about 'human trafficking' so dear to European politicians and stakeholders who taken up this remunerative topic, crossings from Tunisia to Lampedusa, especially on small boats or inflatable dinghies, are often organized by migrants themselves. *Harraga* mostly rely on neighbourhood relations, kin, personal (translocal/national) networks or sheer chance (Mabrouk, 2010). However, *laguatas*, jokers, *khit* (thread), *jellebas* (sheep merchants), mediators, smugglers – i.e. informal border entrepreneurs or travel agents – have a recognized social status. They have a strong incentive for delivering their service to the satisfaction of people who rely on their skills and, therefore, have to maintain, display and enhance their reliability, honesty, sincerity and reputation in a competitive market. Ports are highly controlled by port authorities and the regime's police informers. Local border entrepreneurs work in a highly risky market sector by providing services to migrants who are excluded from the official market and 'legal' mobility. The current fee on the direct route from Tunisia to Lampedusa is around €1,000 and some service providers offer all-inclusive prices covering transfers to final destination.

Routes to Lampedusa via Libya are rather organized and more expensive, as they rely on specialized *passeurs* and smugglers (especially petrol smugglers) who have good contacts to border control and customs officers. In South Tunisia, smuggling provides an important source of income, especially following the international embargo of Libya in the early 1990s. This informal sector went through a severe crisis when the embargo was lifted but they quickly moved into transporting undocumented migrants as one way to create new income.

As local fishermen have boats at their disposal, they also play a decisive role for the organization of *harga*. Because juridical sanctions and administrative restrictions make it difficult to sell boats legally, Tunisian fishermen, who had held the ports during the revolution – often claim that their boats have been stolen (though some were probably hidden sales). On larger boats which are not easily handled by inexperienced *harraga*, the commander (*rais*) and the captain's mate (*dmangi*) are usually former fishermen who can no longer make enough money or who want to diversify.

The dramatic over-fishing of the Mediterranean Sea especially has caused a major crisis in coastal fishing, a crisis that has affected fishermen in Tunisia and also on Lampedusa. The island's local fishing industry is no longer competitive in a globalized market. Structural handicaps, such as high prices for diesel fuel and the lack of efficient transport, have aggravated the situation. Lampedusa is suffering from several conflicts over increasingly scarce resources. Where Lampedusans had once fished along the Tunisian shores, Tunisians are now moving into Italian waters and the island's traditional fishing grounds. This structural crisis has increasingly been the cause of quarrels over the violation of national borders. As the Coast Guard cannot intervene despite the many protests, local fishermen feel that they have been 'assaulted' and 'ruined' by Tunisian fishermen: i.e. they are not recognized as subjects by politicians and public authorities.

There is yet another area where conflicts have arisen. Since the 1980s boom, the tourism industry in Lampedusa is now one of the main economic sectors for its 6,000 inhabitants. As many as 50,000 tourists visit the island each year and give the hotel and restaurant industry its main source of income. Fears that media coverage and visible groups of clandestine migrants ('invisible mobility') could deter tourists (visible and acceptable mobility) are regularly mentioned in populist rhetoric, and not only during local election campaigns.

The economy of camps

Benefiting from the fisherman's traditional hospitality and help for the shipwrecked, the first boatpeople arriving on Lampedusa were cared for of by local volunteers. Everyone remembers that the "*Women cooked for them and collected clothing*", and that they were offered lodgings by the locals, as the former *Carabiniere* officer vividly recalled: '*We didn't know where to lodge them all*'. A group of volunteers, including young people trained by the Italian Red Cross, organized care for them around the clock. Gradually, in line with national legislation on immigration (the *Legge Turco-Napolitano*, 1998), reception centres (*Centri di accoglienza*) were set up and this spontaneous local hospitality became increasingly institutionalized. 'Professionalization' changed the relationship between the new arrivals and the local population, as the many gestures of hospitality were converted into fictions of institutional receptions and best practices. Boatpeople who had been visible as individuals for the locals suddenly became invisible in two ways: as *clandestines* and as 'extraterritorials' who were assigned to a segregated 'heterotopic' place of transition and a space for 'inclusive exclusion' (Agamben, 1995; Diken and Laustsen, 2005).[12]

The first Centre of Temporary Stay and Assistance (*Centro di Permanenza Temporanea ed Assistenza, CPTA*) on Lampedusa was established in July 1998 and managed by the Misericordia. The former volunteers – who worked with a rather small budget – were replaced and, for the most part, were not offered paid jobs with the new centre. "*Just a small number of us got a job, the director came from Corleone – what else do you want to know?*" a former activist told me using Corleone as a metaphor for mafia and clientelism.

12 "Refugees are marked out by their precise lack of rights. Their a- or extra-territorial form of existence seems to consign them to an abject condition of speechlessness which leaves them with little or no remit to challenge often ill-intentioned depictions (as well as occasional brutality or violence)" (Rajaram and Gundy-Warr, 2004: 37). However, local actors took an active role in the revolt of inmates in 2009, when boat people forced open the gates of the overcrowded facility to protest against their treatment and imminent deportation, and the facility was burnt down. Chanting 'Freedom! Freedom!' and 'Grazie Lampedusa', they joined the residents who were demonstrating and protesting against plans by the government and the Minister of the Interior, Maroni (*Lega Nord*) to convert the centre into a centre of identification and expulsion (CIE). The protest opened up a shared space for protests and the borderlands and the limitations of hospitality became an opportunity for commitment, solidarity, and antagonistic transnational political practices (see Friese, 2011). Local protests in 2011 – the facility was once again burnt down and now remains closed – focused on the lack of participation and a political decision to produce images of disaster and that would force the EU to find a solution.

The former centre was located near the airport at the edge of town. It was made up of several prefabricated houses that could accommodate 150 people, including a refectory, a section for bathrooms, an office wing for the administration, the *Carabinieri* and collaborators of Misericordia who ran the centre until spring 2007. The complex was surrounded by barb-wire and inmates were forbidden from moving freely around the island. After a period (during which they were given almost no information about asylum procedures), boatpeople were 'distributed' by plane to other facilities in Sicily or mainland Italy or deported to Libya.

Due to severe overcrowding, these facilities had to be expanded. During construction of the new centre, fierce local conflicts arose over the designated site. One local faction favoured a new building (in a rather obvious attempt to sell land), while the opposition insisted on using existing barracks, even though this would cause a minor environmental impact. Due to these public protests (known locally as 'strike' (*sciopero*)), the construction of a new building was prevented. The new facility, which was expected to be a "model for all existing facilities" was formally illegal as it was built in a protected area. It was designed to take a maximum of 800 internees and inaugurated in August 2007.

In 2007, two cooperatives (Blu Coop from Agrigento and Sisifo from Catania) associated with Legacoop, a left-wing trade union, formed a consortium called Lampedusa Accoglienza which then won the public contract for running the new centre. Since the Tangentopoli campaign,[13] around 40% of public commissions have been handed to consortiums. Given that participation in the procedure requires a level of legal and financial resources that small cooperatives can rarely afford, several individual cooperatives usually form a consortium with Legacoop serving as an umbrella for submitting tenders for public contracts. However, this outcome dissatisfied not only antiracist groups. Opponents of Legacoop's regional office claimed that Blu Coop and Sisifo had already sealed the deal before the call for tender was published and thus regional competitors had been excluded. In addition, both cooperatives had a reputation for being involved in political parties and cronyism and run by 'family and friends'. Although the tender procedures were legally correct, this affair sheds light on how reception of clandestine migrants has become a business in which personal interests intertwine with political networks and local government authorities.

After publication of the public contract in the newspaper *Manifesto* (1.4.2007), anti-racist groups occupied Legacoop's offices in Bologna

[13] The anti-corruption movement that shook the foundations of the Italian political system in the early 1990s.

and Palermo, calling for cooperatives to respect the tradition and values of the trade union and workers' movements and therefore should not be involved in the detention system. They also criticized the cooperatives for winning the public contract by undercutting the competitor, Misericordia which had previously run the facility, by more than 30%. In order to make financial transactions more transparent and avoid further waste of public resources, the Ministry of the Interior limited the expenses for every 'guest' (*ospite*) to €50 per day. Misericordia bid €50, another Sicilian cooperative, Connecting People, €37, and Lampedusa Accoglienza €33. Sisifo's vice-president, a member of La Margherita, a political party which has since merged into the Democratic Party (PD), stated that they had planned an annual budget of around €2 million, that they would do a marvelous job and that service would not suffer. However, the centre was no longer a centre of temporary stay (CPT) but was soon converted into a centre for aid and first reception (*Centro di Soccorso e Prima Accoglienza, CSPA*), which means, at least in theory, that new arrivals were not expected to stay for more than 48 hours during which they could be identified and could make an application for asylum. In practice, however, inmates have to stay for quite extended periods ranging from 8 to 25 days.[14] The facility employs a staff of around 200 local people and thus contributes to the island's economy. Although local residents complain about the militarization of the island, the local hotels are used to house a regular number of security staff providing them and the managers of restaurants and bars additional revenues beyond the off-peak season.

Financing the centres is a particularly serious issue. The first project Praesidium I (Potenziamento dell'accoglienza rispetto ai flussi migratori) was funded by the EU ARGO programme in 2005 and this was renewed every year since then.

> On 1 July 2005 the European Commission allocated 2 million euros (just under one-third of its 6.7 million euro budget) under the ARGO Work Programme to "support operational activities to address the emergencies caused by illegal immigration in the Mediterranean" (Hamood, 2006: 75).

In 2006, the Ministry of the Interior requested around €760,000 from the European Commission (EC ARGO, 2006a, b). The project is a collaboration jointly run by IOM, the Italian Red Cross, and UNHCR on Lampedusa. Since April 2008 the NGO Save the Children is responsible for dealing with unaccompanied minors. The project's aim is to improve

[14] There are four types of centres: first aid and assistance centres (CSPA), reception centres (CPA), centres for asylum seekers (CARA) and centres for identification and expulsion (CIE). The cost of these centres amounts to €200 per day and the overall budget for the period 1999 to 2011 was €985.4 m (Manconi and Anastasia, 2012).

reception and services for "unaccompanied minors, streamline procedures for identification of migrants, and reduce clashes between different ethnic groups". Praesidium demonstrated that public-private contracts could be run successfully with its fifth year in April 2010 (Ministero dell'Interno, 2010). In 2011, *Save the Children*'s annual budget included an allocation of €52,348 for the Lampedusa emergency (Save the Children, 2011: 38).

Dumping boats

With the arrival of boat people on Lampedusa, another problem has emerged: the disposal of vessels. The boats used by *harraga* to cross the Mediterranean are inevitably making their last journey. Ships and inflatable dinghies are immediately confiscated by the border and customs service (Guardia di Finanza, GdF) and kept at the local dump or at the former US Loran base. In 2007, the municipality's triennial budget allocated €550,000 for the "establishment of a provisional depot for the nautical relicts of the extracommunitarians" funded by the Civil Defense Department (Protezione Civile). Ironically the site, which was designed to take 30 boats, was then closed because its construction had not been authorized and no ecological impact report had been submitted. Incidentally, it was located by a road with a picturesque panorama.

Once a state of emergency was declared for Lampedusa, the Civil Defence Department, as the responsible agency for national and natural disasters, joined the border control regime and took an interest in the local economy. In 2004, it was charged with handling the dumped boats as an 'emergency', in order to circumvent rules on public commissions.[15] A new site was found for the boat dump and Edilmeccanica G. Campione S.r.l. in Agrigento won the public tender for its construction. Edilmeccanica is part of a group controlled by Giuseppe Campione, a politician "rumored to be involved in Mafia activities" (Senato della Repubblica, 2004). However, as the site was badly constructed, some of its walls crumbled in 2006. In 2007, another company, SEAP, was given a provisional commission for scrapping the boats. This contract was revoked by Civil Defense in 2008 but SEAP appealed this decision and succeeded in having it overturned in December 2009. In December 2008, the site was burnt down – though this was probably deliberate, since fires are often used by the Mafia as a Mafia as warning and/or an easy way to avoid paying a high price for ecological scrapping methods. In April 2009, a public tender worth €350,000 was issued (Dipartimento della Protezione Civile, 2009). In the meantime, SEAP's

15 After various scandals and a national outcry, the head of the Civil Defence Department, Guido Bertolaso, was removed from office.

Managing Director,[16] accused the mayor of Lampedusa of corruption for receiving bribes amounting to €70,000. The mayor was taken into custody in July 2009 and removed from office but reinstated after his release in August. He was not reelected as mayor at the next elections in spring 2012.

The 'boat cemetery' is a very impressive site (Friese, 2007, 2011a). In 2007, it was estimated that more than 200 boats and inflatable boats were piled up on the site. Locals wondered why boats were not sold at public auctions as in Malta: "*I could never afford a boat. I would be happy to get even a small one for €2,000 or €3,000 and here new boats are left to rot*", a worker told me. There were also suspicions about the whereabouts of the outboard motors "*They divide them among themselves*", he thought, "*I've never seen any up here. There is a black market, but what can we do? The laws are made by those who rule...*" Even if the motors are neatly piled up in the Loran base, some dump sites are open to the public and costly propellers and other replacement parts are dismantled and disappear in the informal market.

In late September 2011, yet another public tender, worth €1,671,740 was published for the management of boats arriving after the Tunisian revolution (Dipartimento della Protezione Civile, 2011). The declaration of the state of emergency became an ordinary method for processing people who claimed their universal right to mobility. It became as well a prevailing mode to circumvent usual norms of public tenders and part of the political economy of borders.

Catastrophes and the state of emergency

In February 2011, Italy declared a "state of humanitarian emergency in the territory of North Africa [sic] ... in order to allow efficient measures hindering the exceptional influx of aliens on national territory" (Presidenza del Consiglio dei Ministri, 2011). Following suggestions of the Head of the Civil Protection and the Commissioner for the 'immigration emergency', the sum of €300 million was allocated in January 2012 to cope with the "humanitarian emergency due to the exceptional influx of Northern Africans to Italy".[17]

Included in the emergency provisions was a fiscal moratorium for local tourist operators and hotel owners who complained about difficul-

---

16 Also a personal friend of the former Minister of Justice, Angelino Alfano, who is a member of the Popolo della Libertà and from Agrigento.

17 "€101 million were allocated as part of the agreement with Tunisia as stipulated in April 2011, €197.4 million for necessary measures to cope with emergencies and finally €8 million for allowances for the police force and the fire department responsible for ensuring public law and order" (Lettera43, 2012).

ties in their business due to the arrival of migrants and an extensive media coverage which together harmed the island's image as an uncontaminated paradise. Indeed, the dominant media-gaze fostered the social *imaginaire* of being assaulted. The production of these images was part of a political strategy of the former minister of the Interior, Maroni (*Lega Nord*), in order to persuade the EU to provide financial aid. On their arrival, the *harraga* were not immediately transported to Sicily, but had to camp in the streets of the small village. For weeks, the island looked like an open-air TV studio where journalists and freelance photographers from all over the world co-produced a humanitarian crisis. The fiscal moratorium was authorized by Silvio Berlusconi during his visit to the island during this state of emergency, and promised the inhabitants milk and honey. It was extended by the Monti government until June 2012. In July, the newly elected mayor of Lampedusa (a left-wing activist) requested another extension until December, 2012, referring to measures adopted by the government in the aftermath of the devastating earthquake in Aquila. The state of emergency has become the norm and it fosters the business interests of local actors.

A permanent state of emergency and normalization of the exception are promoted by use of highly symbolic images and – paradoxically – by contrasting attitudes towards undocumented migrants. On one hand, in film clips made with mobile phones and widely distributed via *YouTube* display, the *harriga* publish images showing their liberation and agency despite deprivation of their freedom of movement (Friese, 2012). Meanwhile, the dominant media coverage re-produces images of 'biblical exodus' and an alarming 'human tsunami'. These images have thus become a powerful resource which is increasingly used to give legitimacy and confirm the need for cross-border policies and the permanent state of emergency. The depiction of mobile people as a sudden and unexpected catastrophe, similar to such disasters as devastating earthquakes, have influenced the civil society and contributed to a siege mentality, which must then be pacified by efficient and well-calculated measures of border management. These three elements – dominance of mass media images, border management policies and a state of exception – are interrelated: the images confirm the exception, just as the exception needs these images to obtain public and political legitimacy.

As we have seen in recent years, new players have now entered the nascent migration industry. According to IOM's report on *Engaging Business and Civil Society* (2006), the border regime covers a variety of stakeholders and institutional actors from the public and private sector and its vital economic interests. At the intersection between formal and informal economy, it includes entrepreneurs in border zones arranging the voyages of those who do insist on their right to mobility. It encompasses conflicts about resources, the production and commercialization of knowledge and services, media coverage of catastrophe and political discourse. The production of knowledge and images, policy and economic interests coincide.

The reception of *harraga* has become an integral part of the political arena and of powerful networks which bring Lampedusa into contact with the regional and national political context. Thanks to government policies on border management, it has become a lucrative business sector, offering opportunities for public works and commissions, subsidies, jobs, and control of public resources. The border regime has become an entangled web of business interests, new modes of governance, knowledge production and media images. Research on the limits of hospitality and the nascent migration industry not only contributes to the production of knowledge, it also participates in the border regime.

References

Agamben G 1995. *Homo sacer: Il potere sovrano e la nuda vita.* Torino, Einaudi.

Amnesty International. 2005. *Lampedusa: ingresso vietato. Le deportazioni degli stranieri dall'Italia alla Libia* (prefazione di G. M. Bellu). Amnesty International, sezione Italiana, Roma, EGA editore.

Barry A., T. Osborne and N.S. Rose (eds.). 1996. *Foucault and Political Reason. Liberalism, Neo-Liberalism, and Rationalities of Government.* Chicago: Chigaco University Press.

Bauböck R. 1994. *Transnational Citizenship. Membership and Rights in International Migration.* Aldershot: Edward Elgar.

Ben Cheïkh F. and H. Chekir. 2008. "La migration irrégulière dans le contexte juridique tunisien". *Notes d'analyse et de synthèse – Série sur la migration irrégulière module juridique.* CARIM-AS 2008/64. Fiesole: Institut Universitaire Européen – Robert Schuman Centre for Advanced Studies.

Benhabib S. 2004. *The Rights of Others: Aliens, Residents and Citizens.* Cambridge: Cambridge University Press.

Caillault C. 2012. "The implementation of coherent migration management through IOM programs in Morocco", in M. Geiger and A. Pécoud (eds.), *The New Politics of International Mobility Migration Management and its Discontents.* Institut für Migrationsforschung und Interkulturelle Studien (IMIS) Universität Osnabrück, Osnabrück, 133-56.

Carens J. 1995. "Aliens and citizens: the case for open borders", in R. Beiner (ed.), *Theorizing Citizenship.* Albany: Suny Press, 229-55.

De Genova N.P. 2002. "Migrant 'illegality' and deportability in everyday life". *Annual Review of Anthropology* 31: 419-47.

Derrida J. and A. Dufourmantelle. 1997. *De l'hospitalité.* Paris: Calmann-Lévy.

Diken B. and C.B. Laustsen. 2005. *The Culture of Exception: Sociology Facing the Camp*, London/New York: Routledge.

Dipartimento della Protezione Civile. 2009. Bando di gara (30.3.2009), http://protezionecivile.it/cms/attach/bando_di_gara.pdf (Accessed 25.5.2010).

Dipartimento della Protezione Civile. 2011. "Bando a procedura aperta per l'affidamento del servizio di rimozione, trasporto, demolizione e avvio a recupero/smaltimento di imbarcazioni situate presso la ex Base Loran a Lampedusa". 27.9.2011, http://www.protezionecivile.gov.it/jcms/it/view_avb.wp?contentId=AVB14449 (Accessed 4.10.2012).

Düvell F. 2008. "Transit, migration and politics: Trends and constructions on the fringes of Europe". Summary paper prepared for the EU Network on International Migration, Integration and Social Cohesion (IMISCOE) and COMPAS. http://www.compas.ox.ac.uk/fileadmin/files/Publications/Reports/Duvell%20transit%20IMISCOE%20report.pdf (Accessed 4.10.2012).

European Commission ARGO. 2006a. *External borders, asylum, visas and immigration* No. JLS/2006/ARGO/GC/09, http://ec.europa.eu/justice_home/funding_2004_2007_argo/doc/projects_selected_details_2006_en.pdf.

European Commission ARGO. 2006b. *External borders, asylum, visas and immigration* No. JLS/2006/ARGO/GC/10, http:// ec.europa.eu/justice_home/funding/2004_2007/argo/funding_argo_en.htm.

Fargues P. 2009. "Irregularity as Normality among Immigrants South and East of the Mediterranean". *Analytic and Synthetic Notes – Irregular Migration. Series Background paper* CARIM-AS 2009/02. Fiesole, Institut Universitaire Européen – Robert Schuman Centre for Advanced Studies.

Foucault M. 1994. "Des espaces autres", in *Dits et écrits*, Vol. IV, Paris: Gallimard: 752-62.

Foucault M. 2004. *Sécurité, territoire et population.* Paris: Seuil.

Friese H. 2004. "Spaces of hospitality". *Angelaki. Journal of the Theoretical Humanities*, 9(2): 67-79.

Friese H. 2007. *Partire*. Photoessay, http://www.hfriese.de/heidrun_friese/photoessays.html.

Friese H. 2009. "The limits of hospitality", in M. Dikec, N.H. Clark and C. Barnett (eds.), Extending Hospitality: Giving Space, Taking Time, *Paragraph* 32, 1: 51-68.

Friese H. 2010. "The limits of hospitality. Political philosophy, undocumented migration and the local arena". in H. Friese and S. Mezzadra (eds.) special issue, *European Journal of Social Theory* 10, 3: 299-33.

Friese H. 2011a. Photoessay, http://www.hfriese.de/heidrun_friese/photoessays.html.

Friese H. 2011b. "The limits of hospitality. Undocumented migration and the local arena. The case of Lampedusa", in M. Baumann, A. Lorenz and K. Rosenow (eds.), *Crossing and Controlling Borders – Immigration Poli-*

cies and their Impact on Migrants' Journeys. Opladen: Budrich UniPress: 249-72.

Friese H. 2012. "Ya l'babour, ya mon amour – Raï, Rap und undokumentierte Mobilität", in M. Dietrich and M. Seeliger (eds.) *Deutscher Gangsta-Rap. Sozial- und kulturwissenschaftliche Beiträge zu einem Pop-Phänomen.* Bielefeld, transcript: 231-84.

Frontex. 2007. "Frontex-led EU illegal immigration technical missions to Libya, 28 May-5 June 2007", http://www.statewatch.org/news/2007/oct/eu-libya-frontex-report.pdf (Accessed 4.10.2012).

Frontex. 2011. *Press Pack* http://www.frontex.europa.eu/assets/Media_centre/Frontex_Press_Pack.pdf (Accessed 4.10.2012).

Hamood S. 2006. "African transit migration through Libya to Europe: The human cost". *Forced Migration and Refugee Studies*, The University of Cairo, http://www.migreurop.org/IMG/pdf/hamood-libya.pdf (Accessed 4.10.2012).

Honig, B. 2001. *Democracy and the Foreigner.* Princeton: Princeton University Press.

IOM. 2006. "Partnerships in migration: Engaging business and civil society". Power-Point presentation on 'Migration Stakeholders', http://www.iom.int/jahia/webdav/site/myjahiasite/shared/shared/mainsite/microsites/IDM/sessions/92nd/partnerships_in_migration_ppt.pdf (Accessed 4.10.2012).

IOM. 2011. "Cooperation to stabilize at-risk communities and enhance migration management to support transitions in Egypt, Tunisia and Libya", 22.12.2011, http://www.iom.int/jahia/Jahia/media/press-briefing-notes/pbnAF/cache/offonce/lang/en?entryId=31118 (Accessed 4.10.2012).

Kasparek B. and F. Wagner. 2012. "Local border regimes or a homogeneous external border? The case of the European Union's border agency Frontex", in M. Geiger and A. Pécoud (eds.), *The New Politics of International Mobility Migration Management and its Discontents.* Institut für Migrationsforschung und Interkulturelle Studien (IMIS) Universität Osnabrück, Osnabrück, 173-92, http://www.imis.uni-osnabrueck.de/pdffiles/imis40.pdf.

Lettera43. 2012. "Immigrazione, il governo ha stanziato 300 milioni. Fondi per l'emergenza umanitaria legata ai flussi dal Nord Africa". Martedì, 10 Gennaio 2012, http://www.lettera43.it/attualita/35997/immigrazione-il-governo-ha-stanziato-300-milioni.htm.

Mabrouk M. 2010. *Voiles et sel. Culture, foyers et organisation de la migration clandestine en Tunisie*, Tunis: Les Éditions Sahar.

Manconi L. and S. Anastasia (eds.) 2012. *Lampedusa non è un isola. Profughi e migranti alle porte dell'Italia*, http://download.repubblica.it/pdf/2012/Lampedusa.pdf (Accessed 4.10.2012).

Mezzadra S. 2006. *Diritto di fuga. Migrazioni, cittadinanza, globalizzazione*, Verona, Ombre corte.

Ministero dell'Interno. 2007. *Communicati stampa* 29.12.2007, http://www.interno.it/mininterno/export/sites/default/it/sezioni/sal... Tripoli_accordo_per_il_pattugliamento_congiunto.html_1562036151.html (Accessed 4.10.2012).

Ministero dell'Interno. 2007. *Comunicati Stampa, L'accordo Italia-Libia sul pattugliamento: i contenuti*, http:// interno.it/mininterno/export/sites/default/it/sezioni/sala_stampa/comunicati/0866_2007_12_29_I_contenuti_dellxaccordo_con_la_Libia.html (Accessed 4.10.2012).

Ministero dell'Interno. 2010. "Notizie, Immigrazione. Rinnovato per il 5° anno *Praesidium*, il progetto per l'accoglienza", 01.04.2010, http://www.interno.it/mininterno/export/sites/default/it/sezioni...ge%3D16%26text%3Dlampedusa%26start%3D%26end%3D%26type%3Dgeneric (Accessed 4.10.2012).

Ministero dell'Interno Italiano. 2012. "Processo verbale della riunione tra il Ministro dell'Interno delle Repubblica Italiana ed il Ministro dell'Interno della Libia" (3.4.2012), http://www3.lastampa.it/cronache/sezioni/articolo/lstp/458843/ (Accessed 4.10.2012).

Presidenza del Consiglio dei Ministri. 2011. "Dichiarazione dello stato di emergenza umanitaria nel territorio del Nord Africa per consentire un efficace contrasto all'eccezionale afflusso di cittadini extracomunitari nel territorio nazionale", *Gazzetta Ufficiale* No. 83, 11.4.2011, http://www.protezionecivile.gov.it/jcms/it/view_prov.wp?toptab=2&contentId=LEG24032#top-content (Accessed 4.10.2012).

Rajaram P.K. and C. Grundy-Warr. 2004. "The irregular migrant as *Homo Sacer*: Migration and detention in Australia, Malaysia, and Thailand', *International Migration* 42, 1:33-64, http://onlinelibrary.wiley.com/doi/10.1111/imig.2004.42.issue-1/issuetoc (Accessed 4.10.2012).

Save the Children. 2011. *Bilancio al 31 Dicembre 2011*, http://images.savethechildren.it/f/download/bilancio/20/2011_bilancio.pdf (Accessed 4.10.2012).

Sciortino G. 2004. "Between Phantoms and Necessary Evils. Some Critical Points in the Study of Irregular Migration to Western Europe". *IMIS Beiträge. Migration and the Regulation of Social Integration* 24: 17-43.

Senato della Repubblica. 2004. Legislatura 14° – Aula – Resoconto stenografico della seduta No. 686 (02.11.2004) Interrogazioni, http://www.senato.it/japp/bgt/showdoc/frame.jsp?tipodoc=Resaula&leg=14&id=120892 (Accessed 4.10.2012).

Tsianos V. (n.d.). "Die Karte Europas und die Ströme der Migration". *Grundrisse. Zeitschrift für linke Theorie und Debatte*, http://www.grundrisse.net/grundrisse27/stroemeDerMigration.htm (Accessed 4.10.2012).

Walters W. 2006. "Border/Control". *European Journal of Social Theory* 9(2): 187-203.

"Maid in Lebanon"[1]

Lebanon and the World Domestic Service Market

Assaf DAHDAH

Since the 1990s, the international domestic service market has become more complex and extensive with the emergence of "global care chains" (Ehrenreich and Hochschild, 2003). It is experiencing tremendous growth and demonstrates thus both the cause and the effect of national migration policies in a context of growing regional and global inequality (Cox, 2006), demographic and social change, and changing job markets which has generated a new *international lumpenproletariat* (Roulleau-Berger, 2010) with the majority of migrants now women (Frederici, 2002). The migration of female workers, now an integral factor in economic policy in many countries, is the first phase in the development of a world domestic service market in which there are clearly identifiable 'sending' and 'receiving' countries. Of the latter, the Middle East is the main source of demand in domestic services (Cheng, 2006).

Already in the 1970s, Lebanon was importing female labour, mainly from the Seychelles and Sri Lanka, to work as domestic servants (Jureidini, 2003). During the 1970s and 1980s, despite the constant presence of war, migrant networks began to diversify and the number of migrants increased (Bret, 2007). With the return to a relatively stable political situation in the 1990s, on the one hand, and the development of a national and world market for domestic services, on the other, there emerged a new partnership bringing together the Lebanese government, specialized recruitment agencies, and employers. Migrant women were the last link in the chain, once management of the market was delegated to private operators keen to develop their enterprises, but all too often with scant regard for fundamental human rights.

Although men continued to migrate, their numbers were much reduced compared to those of women. This was due to a combination of political policy, market pressure, and personal and family strategies.

[1] Title of a film directed by Carole Mansor (2005) in collaboration with Caritas, the International Labour Organization and the Dutch Embassy in Beirut, see URL: http://www.youtube.com/watch?v=3rY91LCyY4s.

Indeed, few men were willing to accept domestic work and they were rarely wanted by employers. Specific regulations were usually drawn up for this group of migrants with rules on the work and life of thousands of women whose legal status was based on a restrictive contract that gave them only limited rights and mobility.

Lebanon currently has a population of nearly 4 million (CAS, 2009), including more than 200,000 migrant women (Dahdah, 2010).[2] As in other Gulf nations, labour immigration is subject to the *kafala* system. This system transfers responsibility for foreign workers to their Lebanese employer (a *kafil*), without whom they cannot obtain either a residence permit (*iqami*) or a work permit (*ijazete 'amal*). There are a number of specific conditions relating to domestic services which provide a very strict geographic, social and administrative framework ruled by the triad of government authorities, recruitment agencies and employers but few provisions or structures for protecting domestic workers' rights. As a result, a profitable 'migration tunnel' has developed, in other words, a situation in which migrants have very little freedom of action while living within a highly regulated situation. Despite this coercive system of control, thousands of migrant women currently live in Lebanon without work papers and the opportunities of work in the black market are increasing, sometimes encouraged by recruitment agencies and by employers working for government institutions.

Various local and national actors are involved in the public-private partnership that regulates the presence of foreign domestic servants. The 'merchandise' – women – are traded within an international system based on supply and demand in which the important issues have an impact not only on diplomatic relations, the job market and the economy, both in countries of origin and of destination, but also on migrant women themselves. International domestic services stand at the *nexus* gender/migration (Morokvasic, 2008) and development/migration (Reysoo, 2008) and lead to the transnationalization of long-distance mothers (Hondagneu-Sotelo, 1997). At the same time, it is an indicator of a global division of labour (Parreñas, 2001) in which actors nurture, and profit from, weaknesses on both sides of the frontier. When making choices between international domestic service and control of migration, what is the position of the Lebanese government, together with various actors, vis-à-vis a global market for domestic servants in which migrants are mere merchandise?

---

2 Official data provided by the Central Administration of Statistics (CAS) is generally approximate and should be analyzed with caution.

In order to understand how the market for domestic services operates in Lebanon, I will first show that this is not a new market but has in fact been seen many transformations and the emergence of new actors. I will then review how the Lebanese market can provide insights into the economic, social and political stakes involved and the relationship between government and private enterprise, in this instance, recruitment agencies. Finally, this article will illustrate how the market is being redefined and distorted by political parties and in response to resistance from the emerging 'civil society' and women migrants themselves who, despite everything, find ways to work outside the system.

Globalization of the Lebanese domestic services market: the situation changes in the 1970s

Before 1970, only the higher social classes could afford domestic servants and they hired mainly Lebanese and Syrian women from rural areas. Since then, the market has undergone major transformations for a number of reasons: more Lebanese women entering the job market, higher quality of life, the compartmentalization of Lebanese communities, and the civil war and its direct and indirect impact on society (Jureidini, 2003). Gradually, African and Asian nationals have replaced Arab servants as a result of regional and global socio-economic upheavals (Dahdah, 2010).

Initially, during the 1970s, most non-Arab domestic servants came from Seychelles and Sri Lanka (Jureidini, 2003). The civil war in Lebanon (1975-1990) discouraged many migrants, but Sri Lanka remained the main source of servants until the 1990s (Bret, 2007). During the 2000s, the number of migrants increased, with new arrivals of domestic servants coming from the Philippines, Nepal, Bangladesh, West Africa, Madagascar and Ethiopia alongside the *sirlankiyé* (Sri Lankans, the term used for 'maids' in Lebanon).[3] This trend towards greater diversity and more women in the domestic service market emerged once private recruitment agencies became involved. Taking advantage of the absence of government during the civil war, these agencies prospered in a commercial environment which operated without bureaucratic controls. Once the conflicts ended in the 1990s, new structures were introduced for this sector and the market continued to grow. In addition, the new government was ready to support this sector and to benefit from a new source of revenue and, as a result, the number of agencies increased.

---

[3] Figures for Sri Lankan women in Lebanon vary between 15,000 (Ministry of Labour, 2010) and 100,000 (Jouni, 2008). For other nationalities: Philippines, approximately 30,000 (Ministry of Labour, 2010), Ethiopians between 30,000 (Ministry of Labour 2010) and 70,000 (Dahdah, 2010).

According to some sources, there were around 400,000 foreign migrant women living in Lebanon (Cattan, 2009), but others, including the Ministry of Labour and the *Sûreté Générale* (responsible for border controls), the figure was less than 130,000 (CAS, 2010).[4]

In order to understand how the Lebanese market in domestic services has changed in recent decades, we must first take into account the regional context with the development of oil powers in the Gulf and its impact on South Asian countries and the Indian sub-continent.

'Sending countries': the other catalysts for change

With the oil crisis in the 1970s, non-oil producing and non-industrial countries became increasingly poor. Governments therefore decided to introduce policies that encouraged the export of their labour force: in particular, the Philippines and Sri Lanka began to "[train] exportable professionals" (Kofman, 2008) in order to reduce unemployment and encourage inflows of currency through remittances (Bergem, 2006). They turned to specialized agencies as a way of structuring this type of migration: the Philippine Overseas Employment Administration (POEA) and the Sri Lankan Bureau of Foreign Employment (SLBFE) are two examples. South-East Asia sent many contingents of migrants to Asian countries within the region (Singapore and Hong Kong (Noin *et al.*, 1999; Yeoh et Huang, 2000)), Western countries (Mozère, 2010; Scrinzi, 2003) and the Middle East (Bergem, 2006; Cheng, 2006). More than 1.5 million women left the Philippines, Bangladesh, Sri Lanka and Indonesia during the 1990s.[5] In 1995 alone, it is estimated that Asian men and women migrants sent home nearly $US75 billion (approx. €50 billion) in remittances (Boyle, 2002, 536).

Filipino migration offers an example of how structures and methods for organizing international migration were developed and used by government institutions to reduce the growing national debt (Gardiner Barber, 2000). In the late 1980s, the majority of Filipino emigrants were men but in the 1990s this trend was reversed with women representing more than 60% of the 4 million Filipinos living abroad (*ibid.*).[6] According to SLBFE, similar changes took place in Sri Lanka: in 2009, nearly half of 250,000 emigrant workers were female domestic workers, more

---

4 It is difficult to obtain reliable quantitative data on the domestic services market and on migration in general in Lebanon and this makes it difficult to ensure clarity and treatment of information.

5 According to POEA (2010), nearly 100,000 Filipino domestic servants were working abroad in 2010. Of these, nearly 60,000 were living in the Gulf States alone and 30,000 in Hong Kong.

6 In 2009, there were more than 8.5 million Filipinos working abroad (POEA, 2010).

than 100,000 of whom had been hired to work in the Middle East alone (SLBFE, 2009, 11, 23).

Changing perspectives in the Lebanese market

Alessandra Cancedda defined domestic service as follows: "all those services provided by public or private organisations, or by the third sector, which substitute paid work [...] for work which was formerly performed unwaged within the household [...]: childcare, care of the elderly, domestic cleaning (of the house, linen, clothes, etc.), catering, domestic maintenance and gardening" (2001, 19). The international domestic services market is not a new phenomenon: it has existed since the 17th century and the first recruitment agencies were created during the 19th century (Sarti, 2009). The recruitment of workers was already based on a hierarchy of nationalities and professions (Nakanno Glenn, 2009), very similar to today's market for domestic services in Europe and the Middle East (Vaiou, 2002), although in the Middle East 'domestic servants' are often considered to be 'au pairs' (Busch and Cox, 2012; Cuartas Villa, 2012).

Compared to previous periods, the world market for domestic services has gradually grown and become more complex. Domestic services are dominated by international migration and thus contribute to the globalization of labour markets since, in many countries, domestic and health services are heavily dependent on migrants (Kofman, 2008). This is also "the sector offering access to the job market, and indeed, for [millions] of ethnic women, the only professional option available under current legislation" (Moujoud and Falquet, 2010, 181)

Structurally speaking, the domestic services sector has grown because it is based on sexual discrimination, lower wages, racism, and the blurred frontier between legal and illegal work. The combination of all these trends meant that domestic service was less expensive and therefore more affordable, which in turn accentuates further variations in individuals' circumstances (Cox, 2006). Globalization and the spread of neo-liberalism have also contributed to the development of 'low cost' domestic services provided mainly by emigration from poor countries to less poor or richer countries through collaboration in public-private partnerships.

The Lebanese market: a profitable public-private partnership

The domestic service market in Lebanon operates under regulations based on the principle of *kafala*. Under this system, a foreign woman can only enter the country when a request for recruitment has been made by an employer who is required to provide lodgings in their home.[7] On this basis, the woman is given a visa and work permit. Requests must be submitted and processed, a complicated and costly exercise, by a private Lebanese recruitment agency. These agencies have already established contacts with similar agencies in the countries of origin and are thus required to ensure that future employees have completed the various formalities on their arrival. The ministries and customs authorities are responsible *a minima* for providing an administrative structure for these inflows, from which the government receives substantial revenues.

Recruitment agencies: the cornerstone of international networks

The Lebanese government introduced regulations covering the employment of domestic servants in private homes in 1994 (Diab, 2006), followed by the first regulations governing the establishment of recruitment agencies in 2003 and a federation of recruitment agencies, the Syndicate of Owners of Female Housemaid Employment Agencies (SORA, 2011) in 2004.[8] Recruitment agencies are supervised by the Ministry of Labour which is responsible for registering all forms of commerce. The number of authorized agencies has risen from 150 in the early 2000s to more than 400 today (Dahdah, 2010).

In order to obtain accreditation, agencies must first meet the following conditions: payment of $33,000 dollars (approx. €25,000) to the National Housing Bank, a signed and notarized act of conformity with the Ministry of Labour's regulations, and a clean criminal record (Bret, 2011). In 2010, each agency was authorized to bring in 150 women per year and this figure rose to 200 in 2011, but only they must link migrants to specific requests from employers which list the desired characteristics for their future servants: nationality, age, physical characteristics, family situation, and level of education. Agencies are not allowed to build up 'stocks' but, according to Mr M., a Beirut agent interviewed in November 2011, "some agencies, who want to make more money, have set up a system of illegal sponsoring and of 'ready maid'. They bring in women in anticipation of future requests and constitute stocks, even though they are supposed to respond to applications! That is how they manage to meet their quotas so quickly and they ask other agencies to bring women in for them".[9]

7 This is not obligatory for men or other professional sectors open to foreigners.

8 Renamed the Syndicate of Owners of House Workers Recruitment Agencies in Lebanon in 2008, http://sora-lb.com/ [Accessed 9.9.2011]. By removing the reference to women, the Syndicate clearly wanted to change its image and demonstrate its intention to recruit men.

9 All interviews were conducted in Arabic.

Lebanese agents generally work in collaboration with agencies in countries of origin. In 2009, there were more than 700 recruitment agencies in Sri Lanka, mainly in Colombo, some of which send several thousands of workers a year: more than 10,000 for the Lord Manpower Agency alone (SLBFE, 2009). Candidates for emigration pay the agencies fees between $200 and $500 (Jureidini, 2003, 100). These intermediaries, who are often Lebanese expatriates, also receive a fee from Lebanese agencies and this increases the fees they can charge to employers since they are not subject to regulation. Tariffs are market-driven and agencies can fix their prices amongst themselves, adapting their fees to the law of supply and demand and to priorities in terms of ethnic and professional criteria. In other words, according to Mr M., the cost of hiring Filipinas, who have a reputation for being "*more qualified, better educated, cleaner and white*", is higher than for Ethiopian or Sri Lankan women. Such priorities are reflected in the salaries paid to employees: $150-200 per month on average (*Human Rights Watch*, 2010, 33). Indeed, there are stark differences from one nationality to another, with Filipinas, the "Mercedes Benz of servants" (Mozère, 2001), generally receiving the highest wages.

In protest against the poor working conditions offered to their nationals, Ethiopia and the Philippines banned emigration to Lebanon in 2006. However, they were not able to stop women workers from leaving their shores. This has led to a series of negative outcomes: Lebanese actors, including the government, are placed in a difficult position vis-à-vis other countries; migrants are forced to make more complex journeys; and they face more intermediaries in their attempts to circumvent these bans and this obviously increases the cost. According to Mr M., "with all that money, [sending] governments have no reason to open the gates, they are making more money out of the market for clandestine migrants" through a variety of local intermediaries. During interviews with two agents in Beirut, we learnt that, in 2005, an employer would pay $900 to recruit an Ethiopian woman, whereas today the fee has increased to $1800, since $900 must be paid to intermediaries in countries of origin and transit (Sudan, Yemen, or Egypt) before a woman migrant can arrive in Lebanon. A one-way air ticket costs about $300 and the passport, entry visa and obligatory medical examinations required by the Ministry of Labour (AIDS, malaria, tuberculosis) another $400. The Lebanese recruitment agency is left with about $200.[10] In order to reduce their costs, agencies try to establish or develop connections with other countries, most recently with Bangladesh and Kenya, or even Latin America, but the air tickets are too expensive, according to Mr Y.,

---

[10] For Filipina maids, the fee is closer to $4,000 in 2012, compared to $2,000 in 2005.

a Beirut agent interviewed in September 2011. In addition, agencies must respect the Sûreté Générale's regulations which allow women to enter from some countries, but not from others, in particular from Eastern Europe.[11]

The Government in a muddle: how to control frontiers without upsetting a lucrative market

The Lebanese government, which is responsible for border controls, manages the immigration of women workers through an administrative and customs organization from which it receives substantial revenues. Acting for the state, it not only controls and supervises the market, via the Sûreté Générale and the appropriate ministries, but also must ensure that there is a market in domestic service from which it can earn profits through its partnership with recruitment agencies and employers.

To bring foreign women into the country, agencies must obtain the necessary papers including a three-month visa. Once the application file is complete and has been forwarded to the country of origin, the women are flown to Beirut where they are 'welcomed' by the Sûreté Générale which has set up a special service for them: on their arrival, the women are identified and sent to a waiting room where they are supervised by agents until their *kafil* (employer or sponsor) comes to collect them.

The kafil then pays $600 for a one-year work and resident permit which must be renewed by the Ministry of Labour. Renewals of these permits alone provide the government, if we take the official figure of 130,000 women migrants, with revenues of approximately $80 million per year.

In addition to these sums, they must also pay fees for health insurance and deposits. Given that these women migrants do not have the right to work and to social security benefits (like many Lebanese who do not have social security), the Ministry of Labour requires that, in principle, employers take out an insurance policy for an annual fee of $70 to $90. Finally, the kafil has to pay a deposit of $1,000 to the Ministry which will not be repaid until the migrant has left the country for good.

---

[11] Recruitment agencies are forbidden to bring Eastern European women as they are unofficially 'reserved for prostitution', a market not recognized by Lebanon's Sûreté Générale which issues 'artiste visas' for women hired for night clubs and cabarets, cf. http://www.general-security.gov.lb (accessed 3/1/2012).

Reception of women migrants at the Sûreté Générale desk at Beirut Airport (Mohammad Azakir, *The Daily Star*, 9 September 2011)

The *kafala* system operates only in the domestic service market and allows the government to make a profit from its minimal investment because controls of migrant women after their entry into the country are delegated to employers and recruitment agencies.[12] This partnership also includes NGOs as the Sûreté Générale has a convention with Caritas, an association which is responsible for helping migrant women when they find themselves in difficulty.[13] Finally, the government delegates the legal responsibility and supervision of migrant women to their employers who control their daily life while restricting their movement to the homes, sometimes without a day of rest or a room for their own use. To date, the government has not issued any concrete measures,[14] such setting up a social service, for the supervision of domestic work by

12 Agencies are responsible for migrant women for the first three months of their stay in Lebanon, during which period they must offer an 'after-sales service' if the employer has problems with the employee, or vice versa. The agency can decide to send a migrant back to her country if she does not meet the employer's expectations, despite the loss for their business. Often heavily in debt, some women migrants look for ways to stay in Lebanon, cf. the highly publicized case of Alem Dechassa (Jureidini, 2012).

13 Having signed a Memorandum of Understanding, Caritas is allowed to visit women migrants in prisons and provide them with assistance and care.

14 Apart from a hotline allowing migrant women to reach the Ministry of Labour's services.

migrant women. As all these social, economic and health responsibilities have been delegated to employers, the situation of domestic servants continues to deteriorate since many employers prefer to pay the lowest possible wages for a service for which they believe they pay too dearly.

Because of these working conditions, thousands of migrant women flee the homes of their employers. Escape is not in itself an infraction, especially if they have been ill-treated. Migrant women and associations have denounced the fact that police authorities advise employers to file a complaint against their maids, for example for stealing, as this terminates their legal responsibility (Bret, 2011). However, a maid without a *kafil* or in hiding becomes *de jure* an illegal migrant.

Poster appealing for help in finding an Ethiopian migrant in flight in Dora (Beirut suburb) (Photograph: A. Dahdah 2011)

For the government, a migrant woman in flight must be found and, if possible, returned to domestic service. If not, she will be sent to prison and returned to her country of origin. Expenses relating to expulsion are supposed to be paid by the employer who usually refuses. In the event of a court case, if a migrant woman cannot afford to pay the fees, she can ask for help from her national or religious community. However, if her passport has been confiscated by her employer, she must obtain identity papers.[15] Many women remain in Lebanon – voluntarily or involuntarily, because they cannot overcome these administrative and financial difficulties. Some find jobs on the black market and remain fugitives, while others pay a fee of about $1,000 for a ghost sponsor (substitute *kafil*) who is supposed to cover the service and administrative costs that will allow them work free-lance (Moukarbel, 2009). According to interviews with migrant women, some recruitment agents may offer this service, which is illegal, and even take the fees without obtaining the necessary papers for the woman so that she loses her money.

Even for these contentious, and barely legal, situations, the Lebanese government is involved in organizing, assisting, and benefitting from these migrants since it allows them to continue without upsetting the market. Nevertheless, the absence of any real protection means that there are numerous cases of ill-treatment and death and as a result, an increasingly active civil society has now emerged.

Coming out of the maid's room: steps towards reforms in the domestic service market

The system, which should be lucrative and discrete, is in fact caught up in a tangle of individual, collective, national and international issues which have little to do with the Lebanese government's ability to manage the situation. The number of migrant domestic servants in Lebanon and the growing awareness by individuals and associations of the ethical problems involved have generated mounting tension in which the government finds itself playing the role of both judge and defendant. This has led to greater involvement by migrant women who, by their presence and their flouting of the rules, are upsetting the system that was initially set up as a way to reduce them to '*birds of passage*' (Bouchetout, 2007; Morokvasic, 1984).

The Lebanese market: adaptation and pressure

The embargos against the supply of domestic servants by the principal countries of origin have seriously disrupted the organization of migrant domestic services in Lebanon. The decision by Ethiopia and the Philippines to ban the emigration by their nationals to Lebanon has had a snowball effect, encouraging other governments, Nepal and Madagas-

15 Migrants from countries that have a consulate in Lebanon can obtain papers to cover the period until their situation is resolved and they can leave the country. Others find themselves in often inextricable situations. According to *Kafa*, nearly 80% of employers confiscate the passports of their maids to stop them from leaving, with the tacit approval of the authorities. Recuperating their papers can entail long negotiations with their ex-employers.

car, to take similar steps. Governments have officially denounced the low incomes and lack of protection for their workers who are not covered by Lebanese labour legislation. Lebanon is not the only country in this situation as several Gulf States, including Saudi Arabia, have also been denounced by 'supplier' governments for the lack of protection afforded their labour migrants. These denunciations have sometimes led to diplomatic tensions, as experienced by the Philippines and Indonesia (*Now Lebanon*, 2011). According to Mr Y., a Beirut agent, "*the Gulf States have oil and thousands of Filipinos and Ethiopians work there [...]. They cannot prevent migration to those countries. But the stakes are not the same in Lebanon*".[16]

Recruitment agencies, government authorities and migrant women have had to adapt to these bans, which have forced women into taking different routes or travelling in disguise (*L'Orient-Le Jour*, 2011). As a result, many migrant women leave their countries illegally but enter Lebanon legally and are therefore all the more vulnerable. The two agents cited in this article say that they can no longer obtain a translated version of the contract that migrant women must sign because the consular offices no longer acknowledge the arrival of domestic servants. Nevertheless, thousands of Ethiopian, Filipina and Madagascan women are arriving to work in Lebanon only to realize that the situation is totally confused and that they have to accept harsh conditions (El Hage, 2011a). For example, R., an Ethiopian maid working in Beirut who was interviewed in March 2012, stressed that the role of sending countries is also very ambiguous: "*If my government no longer wants women to die, it must stop giving them passports when it sees they have left for Lebanon. But for them, we are just barrels of oil. And why does the Lebanese government let us come into their country when it knows that it is forbidden?!*"

The number of maids in Lebanon continues to grow and their presence has had a profound influence on the typical Lebanese family which is itself going through a major transformation with the growing number of Lebanese women entering the job market (Yaacoub and Badre, 2011). According to Agent Y, the 'maid' is now a necessity for all social classes: "*Here everyone wants to be like everyone else. It doesn't matter whether you have a salary of $1,000 or $5,000! [...] And yet an Ethiopian maid costs on average about $300 per month and that covers everything. But now, regardless of the household's income, all women want a maid in the house. It's become the norm*". Of course, this statement does not take into account free-lance maids and illegal migrant

16 As discussed in the previous article, Saudi Arabia threatens to close its market to Filipinos and Indonesians, in contrast to the situation in Lebanon.

women working in the cleaning, hotel and restaurant sectors instead of Lebanese and Arabs women for whom these jobs are too degrading and above all poorly paid.

The government is unable to enforce its regulations which in any case are now incoherent. They are supposed to be both strict and restrictive in terms of controls on maids' social and physical presence, but are in fact lax in protecting the rights of foreign workers because they delegate responsibility for migrant women to agencies and employers. Finally, the government has a vested interest because the authorities allow a growing number of women to enter the country.

NGOs: new actors upsetting the market

The poor working conditions and often inhuman treatment inflicted on migrant women are now being denounced and publicized by Lebanese associations for the defence of migrants and women. Following a damning report by *Human Rights Watch* (2010, 45),[17] the civil society has taken up the issue by creating collectives and specific actions for the defence of migrants within existing associations. These collectives and associations – Kafa, the AntiRacism Movement, Insan, Migrant Workers Task Force, Nasawiya and EthiopianSuicides – have denounced the *kafala* system and racism in Lebanon with public and online protests.[18]

The associations have contributed to greater media coverage of the problems created by the government's free market approach to the management of foreign workers. These new actors have strongly denounced the government, while other associations, including *Caritas*, try to help migrant women by working with the authorities. *Caritas* is also operating in prisons and at the airport and offers migrant women safe houses where they receive care and lodgings while waiting for a solution to their situation. At the same time, Lebanon is under pressure from other countries to reform its *kafala* system (Bladd, 2010).

The government is thus caught in a trap of its own devising and is trying *mutadis mutandis* to improve working conditions for women migrants without actually changing the *kafala* system or rescinding the obligation to live with employers. The president of *Nasawiya*, in an interview in March 2011, said that "*the government does not look after*

17 According to this report, at least one domestic servant died each week in Lebanon in 2007 and 2008. The authorities usually handed down a verdict of suicide and did not trouble employers further.

18 See campaigns such as *My Srilankan is Lebanese* (http://www.youtube.com/watch?v=0-dtxEO3GjA) and Migrant Workers Task Force's campaign against the Ministry of Tourism's advertisements (http://www.youtube.com/watch?v=8mdc-omQEbo&feature=player_embedded).

the welfare of migrants; it only wants to improve its reputation which has been tarnished by many negative reports". The former Minister for Labour, Boutros Harb, talked about the importance of changing job contracts for domestic servants in order to give employees more rights – in terms of salary, leisure time, health services – and employers greater obligations but did not provide sufficient adequate resources to ensure that these rights were respected. Clearly, the objective is above all to improve Lebanon's reputation with sending countries and the rest of the world: "*the negative (and justified) reports from international organizations about poor treatment of household staff [must not continue] to tarnish the reputation of Lebanon as a country which respects human rights*" (cited in El Hage, 2011b).

The associations are trying to inform the Lebanese population while simultaneously creating a federation of foreign workers through collective actions in a society where migrants are becoming increasingly visible and organized.

From migrant woman to immigrant woman?

The *kafala* system and management of foreign workers in Lebanon are designed to ensure that male and female migrants remain mobile and invisible, by discouraging any desire to settle in the country. It is an economic policy that seeks to keep them certain social and professional categories and limit their physical mobility in order to provide cheap labour for the '3D jobs' (dirty, difficult, and dangerous) (Kawar, 2004, 74). This policy is now being challenged by migrants who are nevertheless settling in Lebanon, finding support structures, developing networks, organizing national and religious communities, and becoming *in fine* increasingly visible actors in society and business in Lebanon.

These circumventions are the outcome of the rigidity in the Lebanese legislation itself which encourages, by imposing a certain number of human, economic, social and spatial constraints, the 'clandesinization' of migrants who prefer to stay and earn a living illegally in Lebanon than risk being expelled. To make matters worse, the legislation does not provide any instruments for preventing this situation, so that the government has lost all control over this situation while trying in vain to curb it with repressive measures. A public servant in the Ministry of the Interior said in an interview in April 2010: "*the government has to control migration but the Lebanese economy depends on the presence of illegal foreign labour*".

In this context of laxity and contradictions, migrants are taking advantage of the absence of governmental authority in some areas of Beirut to rent an apartment, find a better-paid job, and sometimes start a family despite the difficulties, fear and racism they have to face.

Conclusion

The international domestic services sector has changed because of regional and global disparities which have had a strong impact on countries involved, whether as purveyors or as consumers of domestic services in a global market. The Lebanese government is fully implicated in the trade in human resources, as are Middle Eastern countries, through its private-public partnerships with recruitment agencies and employers. The government is presumed to be the guarantor and the beneficiary of the migratory tunnel, which it is obliged to keep open.

Recruitment agencies are the keystone for a system that the government uses to control the mobility of workers and that provides a substantial source of revenue in return. Clearly this situation is extremely ambiguous, and agencies have taken advantage of the government's generosity, and of its concrete and legislative collaboration, in order to profit from the flow of migrants, all too often to the detriment of migrant women.

But we have only looked at the Lebanese end of the tunnel and it is important to take into consideration public and private actors in sending countries: specialist governmental institutions and recruitment agencies hiring candidates for emigration. These agencies take their cut from the expenses incurred in sending migrants abroad and from services rendered to their Lebanese counterparts. Governments, meanwhile, use the system to structure these displacements and encourage the return of currencies through remittances.

Nevertheless, the system has its limits and it is beginning to suffer from its complexity and from the circumventions and resistance by civil actors and by migrant women who are more and more challenging their living conditions. The government does not have the tools for implementing their policy of creating turnover and controlling migrants' social mobility. Because of its long history of migration, there are many loopholes in the system and thousands of migrant women are taking advantage of these to leave temporarily, or even permanently, the migratory tunnel in which they are supposed to remain. Their indocility is facilitated by the networks that they created, sometimes with help from recruitment agencies which profit from their irregular situation. The partial delegation of control to agencies – which are more interested in having a new opportunity for making money – is becoming less effective and offers greater opportunities for circumvention to people who know how to take advantage.

References

Beaugé G. 1986. "La *kafala*: un système de gestion transitoire de la main-d'œuvre et du capital dans les pays du Golfe". *Revue européenne de migrations internationales*, 3: 109-122.

Bergem, K. 2006. "The role of the state in the in-migration of domestic workers to Jordan and the GCC countries". *Migration et politique au Moyen-Orient*. Beyrouth: IFPO: 61-80.

Bladd J. 2010. "US urges GCC states to scrap sponsorship system". *Arabianbusiness.com*, http://www.arabianbusiness.com [Accessed 15.6.2010].

Bouchetout Y. 2007. "Srilanka Land: ou comment devenir visible à Beyrouth", in M. Davie (ed.) *Fonctions, pratiques et figures des espaces publics au Liban.* Beirut: University of Balamand: 321-329.

Boyle P. 2002. "Population geography: transnational women on the move". *Progress in Human Geography*, 26: 531-543.

Bret J. 2007. "Circulations transnationales et travail disqualifié au Moyen-Orient. Les travailleurs non arabes au Liban". *Hommes & migrations*, 1266: 96-107.

Bret J. 2011. *Travail global et production d'un individu servile. Activités économiques et migrations de travailleurs non arabes au Liban.* Lyon: Université Lumière Lyon 2.

Busch N. and R. Cox. 2012. "Latin American Workers in London". Conference Proceedings *Dimensiones Del Empleo Doméstico LatinoAmericano.* May 2012, Monterrey.

Cancedda A. 2001. "Employment in household services". European Foundation for the Improvement of Living and Working Conditions, http://www.eurofound.europa.eu (Accessed: 30.11.2010).

Cattan N. 2009. "Mobilité et territorialité des employées de maison Sri Lankaises à Beyrouth", Conference proceedings, *Genre en mouvement. Conflits, négociations et recompositions.* Paris (unpublished paper).

Central Administration of Statistics (CAS). 2009. *Residents in Lebanon*, http://www.cas.gov.lb [Accessed 23.3.2011].

Cheng S-J. 2006. *Serving the Household and the Nation: Filipina Domestics and the Politics of Identity in Taiwan.* Oxford: Lexington Books.

Cox R. 2006. *The Servant Problem. Domestic Employment in a Global Economy.* New York-London: I.B. Tauris.

Cuartas Villa L. (2012), "Jóvenes Au Pair: Figura invisible de la migración y del empleo doméstico internacional. El caso de Francia y Estados Unidos". Conference proceedings, *Dimensiones Del Empleo Doméstico LatinoAmericano.* May 2012, Monterrey.

Dahdah A. 2010. *Les espaces de l'altérité dans une métropole méditerranéenne: la visibilité des migrantes non arabes dans le Grand Beyrouth*, Masters 2 thesis, Aix-en-Provence, Université de Provence.

Diab N-A. 2006. "Le droit libanais relatif aux migrations internationales". Analysis and Summary. Florence: *CARIM.*

http://www.carim.org/index.php?callContent=364&callCountry=2420 [Accessed: 4.12.2011].

Ehrenreich B. and A. Hochschild. 2003. *Global Woman: Nannies, Maids, and Sex Workers in the New Economy*. New York: Metropolitan Books.

El Hage A-M. 2011a. "Pas de rapatriement en masse des employées de maison malgaches". *L'Orient-Le Jour*, 9.2.2011.

El Hage A-M. 2011b. "La main-d'œuvre domestique étrangère au Liban toujours dans l'attente d'une loi". *L'Orient-le Jour*, 27.5.2011.

Federici S. 2002. "Women, Reproduction and Globalization". Conference proceedings – Yvonne Preiswerk Collection: Geneva.

http://www.graduateinstitute.ch/genre/page2887_fr.html (Accessed: 3.11.2010).

Gardiner Barber P. 2000. "Agency in Philippine women's labour migration and provisional Diaspora". *Women's Studies International Forum*, 23: 399-411.

Hondagneu-Sotelo P. and E. Avila 1997. ""I'm Here, but I'm There". The meanings of Latina transnational motherhood". *Gender & Society*, 11(5): 548-571 (Consulted: 23.1.2012).

Human Rights Watch. 2010. *Without Protection. How the Lebanese Justice System Fails Migrant Domestic Workers*. URL: http://www.hrw.org (Accessed 9.12.2010).

Jouni H. 2008. "Les immigrés illégaux au Liban". Legal approach, Analysis and Summary. Florence: *CARIM*. http://www.carim.org/index.php?callContent=364&call Country=2420 (Accessed: 4.12.2011).

Jureidini R. 2003. "L'échec de la protection de l'État : les domestiques étrangers au Liban". *Revue européenne des migrations internationales*, 19: 95-127.

Jureidini R. 2012. "Lebanon's ways are sponsoring suicide". *The Daily Star*, http://www.dailystar.com.lb [Accessed 10.4.2012].

Kawar M. 2004. "Gender and migration: Why are Women more Vulnerable?" Conference proceedings, Yvonne Preiswerk Collection: Geneva, 71-88. http://www.graduateinstitute.ch/genre/page2865_fr.html (Accessed 3.11.2010).

Kotman E. 2008. "Genre, migrations, reproduction sociale et Welfare State". *Cahiers du CEDREF*, 16: 101-121.

L'Orient-Le Jour. 2011. "De fausses nonnes philippines, trahies par des chaussures rouges, tentaient de gagner le Liban". 31 January 2011.

Morokvasic M. 1984. "Birds of passage are also women". *International Migration Review*, 18: 886-907.

Morokvasic M. 2008. "Femmes et genre dans l'étude des migrations: un regard rétrospectif". *Cahiers du CEDREF*, 16: 33-56.

Moujoud N. and J. Falquet 2010. "Cent ans de sollicitude en France. Domesticité, reproduction sociale, migration et histoire coloniale". *Agone*, 43: 169-195.

Moukarbel N. 2009. *Sri Lankan Housemaids in Lebanon*. Amsterdam: Amsterdam University Press.

Mozère L. 2010. "La mondialisation comme arène de 'trouvailles accumulées'? Des domestiques philippines à Paris", in J. Falquet, H. Hirata, D. Kergoat, B. Labari, and N. Le Feuvre (eds.) *Le sexe de la mondialisation. Genre, classe, race et nouvelle division du travail.* Paris: Sciences Po. Les Presses, 151-164.

Mozère L. 2001. "La Philippine ou la Mercédès Benz des domestiques. Entre archaïsmes et modernité". *Sextant*, 15-16: 297-318.

Nakano Glenn E. 2009. “De la servitude au travail de service: les continuités historiques de la division raciale du travail reproductif payé”, in E. Dorlin and A. Bidet Mordrel (eds.). *Sexe, race, classe, pour une épistémologie de la domination*. Paris: PUF, 21-70.

Noin D., R. Jackson, S. Huang, B. Yeoh, and M. Guillonlien. 1999. “Les migrations internationales des domestiques philippines. Contextes et expériences aux Philippines et à Singapour”. *Revue européenne des migrations internationales*, 15: 37-67.

Now Lebanon. 2011. “Saudi Arabia bans domestic workers from Indonesia, Philippines”. http://nowlebanon.com (Accessed 29.6.2011).

Parreñas R. 2001. *Servants of Globalization. Women, Migration, and Domestic Work*. Stanford: Stanford University Press.

Philippines Overseas Employment Administration (POEA). 2010a. “OFW Deployment by Occupation, Country and Sex – New hires Full Year 2010”. URL: http://www.poea.gov.ph/stats/statistics.html (Accessed 5.1.2012).

Philippines Overseas Employment Administration (POEA) 2010b. “Stock Estimate of Overseas Filipinos”. URL: http://www.poea.gov.ph/stats/statistics.html (Accessed 5.1.2012).

Reysoo F. 2008. “Ambivalent developments of female migration: Cases from Senegal and Lebanon”. In T. Van Naerssen, E. Spaan and A. Zoomers (eds.) *Global Migration and Development*. Routledge, New York.

Roulleau-Berger L. 2010. *Migrer au féminin*, Paris: PUF.

Sarti R. 2009. “La globalisation du service domestique dans une perspective historique, XVIIe-XXe siècles”. *Genre et travail migrant. Mondes atlantiques, XIXe-XXe siècles*. Paris: Publibook: 53-82.

Scrinzi F. 2003. “Les employées de maison migrantes en Italie. Ethnicisation et contrôle dans le circuit catholique à Gênes”. *CWS/cf Canadian Women Studies/les Cahiers de la femme* 22 (3/4):182-187.

Sri Lankan Bureau of Foreign Employment (SLBFE). 2009. *Annual Statistical Report of Foreign Employment*. http://www.slbfe.lk/article.php?article=68 (Accessed 5.1.2012).

Syndicate of Owners of House Workers Recruitment Agencies in Lebanon (SORA), URL: http://sora-lb.com/ (Accessed 9.9.2011).

Vaiou D. 2002. “In the interstices of the City: Albanian women in Athens”. *Espace populations societies*: 373-385.

Yaacoub N. and L. Badre. 2011. *The Labour Market in Lebanon*, Central Administration of Statistics, http://www.cas.gov.lb (Accessed 22.1.2012).

Yeoh B. and S. Huang. 2000. “‘Home’ and ‘away’: Foreign domestic workers and negotiations of diasporic identity in Singapore”. *Women’s Studies International Forum*, 23: 413-429.

The Palestinian Olive Industry and Israel's Separation Policy

Trading at the Border (2002-2012)

Arnaud GARCETTE

This chapter will discuss the way Palestine's olive industry has changed in the West Bank since the construction of the Separation Wall (2002) and the tightening of Israeli controls on Palestinian mobility. It will highlight the different ways that the Wall and the attendant security measures have affected the local population and also study the Palestinians' ability to come to terms with these constraints in developing their professional lives.

Olive production, which represents up to 25% of agricultural GDP and involves nearly 100,000 families, is an important sector for the Palestinian economy. It also acts as a buffer against economic crises, as was seen during Palestine's 're-ruralization' process (Meneley, 2008a) during the first and second Intifada (1987-1993 and 2000-2005 respectively) and during the Palestinian Authority's budget crisis in 2006-2007 (Mansouri, 2007).

This sector has not attracted much interest from researchers in recent years, even if a number of articles have focused on the olive tree as an emblem and as a political issue as a result of the Israeli-Palestine conflict (Pirinoli, 2005; Abufarha, 2008; Braverman, 2009). It is very often presented as the symbol of injustice and of the way the Palestinian community has become deracinated because of the occupation, even if we have still much to learn about the full impact of these changes on the community and its economy. For example, Bernard Botiveau and Edouard Conte have explained that, in this research market, "we will see a multitude of images showing the destruction of olive trees, but this will not tell us much about how the olive market and the structure of Palestinian communities have been transformed" (Botiveau and Conte, 2005:13). It is difficult to know how to minimize the dramatic and highly publicized impact of confrontations between Israeli soldiers and settlers and Palestinian olive farmers: for example, 1,639,030 trees were destroyed in the Palestinian Territories, including 547,996 olive trees, between 2000 and 2008 (PMoA, 2008). Nevertheless, if we limited our discussion to this simple statement, we could find ourselves overestimating for the occupying power, while underestimating the Palestinians' ability to bounce back and to overcome, or sometimes exploit, the control and separation mechanisms that have been established.

In reviewing changes in olive production methods imposed by the separation policy, we will seek to analyze 'the separation economy' by taking into consideration relations among the various actors in the Israeli-Palestinian region (Natsheh and Parizot, 2011; Parizot, 2006; Garb, 2011; Grinberg, 2011). We will also identify the many ways in which the Wall and its security systems have affected the Palestinian economy and examine the Palestinians' ability to circumvent these constraints in order to maintain their professional activities. The objective of this article is therefore to look at what lies behind specific events affecting olive production in Palestine. It will focus on these sectorial changes which should not be seen as merely influenced by constraints. We must also examine the ways that the actors themselves have introduced changes by re-orienting and restructuring their activities because of these constraints.

I will first review the impact of a wide range of Israeli controls on the development of olive production in the West Bank. I will then demonstrate that this policy of confinement that upset the Palestinians' ability to manage their movements and their working day as well as the management of their agricultural tasks from day to day. Finally, I will then demonstrate how olive farmers overcome these constraints by involving other actors and review the social and economic impact of their intervention.

This research is based on two years of fieldwork undertaken as part of my PhD research (January 2010-December 2011). In addition, it takes into account not only a rich collection of articles but also observations obtained through a series of interviews with actors involved in this sector, from olive growers to retail outlets, covering all aspects including harvesting and processing for the market.

Containment of Palestinians: an obstacle for the development of olive production

Analyses of the impact of restricted mobility have often focused on the many physical obstacles placed on roads in the West Bank without revealing the profound repercussions that they have for transporting and selling olive oil.[1] The fragmentation of Palestinian territory caused by

---

1 With the exception of Yaakov Garb (2008).

the presence of these obstacles has certainly slowed the emergence of dynamic initiatives by individual and collective enterprise.

Developing Palestinian structures for cultivation and marketing in response to confinement

The best-known manifestation – due to the high level of media coverage – of Israel's control systems is the Separation Wall, the first sections of which were installed in 2002. Its perimeter will measure more than 700 kilometres and will make profound and long-term changes to the map of the Palestinian Territories. On completion, approximately 85% of the Wall will be located inside the West Bank, cutting nearly 10% of Palestinian territory into enclaves containing many of Palestine's wells and most of its fertile agricultural land.

The Wall disrupts farmers' ability to visit their lands if they are within the seam zone, as the area between the Wall and the Green Line[2] is known,[3] where they have planted nearly a million olive trees. Before they can work in their olive orchards, farmers must obtain a visitor's permit from the Israeli military authorities. In 2011, 42% of these applications were rejected, slightly more than in the previous year (39%) and, in some regions, this percentage is much higher: in the northern West Bank, for example, only 50% of applications were granted (UNOCHA 2012). In addition, the conditions for obtaining these permits are particularly stringent and applicants must meet strict security criteria imposed by Israel (OXFAM 2010). The criteria are not clear and little justification is provided for rejections, yet unauthorized entry into the seam zone is punishable by five years' imprisonment. Applicants are also required to produce property titles for the farmland in question, even though the transfer of land from one generation to another is rarely accompanied by title documents in the Palestinian Territories. As a result, permit-holders are often very old, or even deceased, patriarchs. In addition, farmers have had to adapt their production methods because there are strict limits on the equipment they can take across at agricultural checkpoints.

Many farmers decide not to apply for a visitor's permit, either because they are discouraged or they reject the process on principle, especially those living in the Governorate of Jerusalem: Beit Surik, Beit Hanina, Kharayib Umm al-Lahem. In fact, 60% of families owning land in the seam zone no longer have access to their property since the construction of the Wall (World Bank, 2008). As Ariel Handel has

---

2 This represents 10% of olive trees in the West Bank, according to an official at the Ministry of Agriculture, at Ramallah (interview dated 12 March 2011).

3 Former armistice line between Israel and Jordan from 1949 to 1967.

pointed out, we can see that checkpoints and other control systems have been put in place not so much to regulate movement as to slow down and limit the movement of Palestinians (Handel, 2011). This policy of 'spaciocide' (Hanafi, 2005) resembles a game of Go, according to Sylvie Fouet (2002: 25), where the objective is not to eliminate the adversary but to immobilize him by blocking all the key points in the matrix.

Control systems also have a direct impact on commercial practices. They create barriers against trading in and through Israel and thus prevent olive producers from maintaining access to their principal consumer markets, Palestinian residents in Israel and the Gaza Strip. Controls also have a direct impact on the market value of Palestinian products in two ways. First, opportunities for economies of scale are reduced: Palestinian farmers are only able to generate mediocre profits because the amount of land they can cultivate is limited and cannot be increased (World Bank, 2008). In addition, their sales outlets are scattered (FAO, 2009): in 2007, Palestinian drivers were banned from using more than 700 km of roads (World Bank, 2007). Second, these difficulties increase production and delivery costs.[4] There are only six checkpoints for delivering merchandise to Israel and checking procedures are complex and costly (Garb, 2008). Higher charges required for ensuring delivery and export of their produce clearly increase their production costs: for example, diversion of regular delivery routes to designated checkpoints; additional 'security' restrictions imposed by Israel on warehousing goods and loading trucks;[5] delays caused by queues at checkpoints and for security inspections. Furthermore, World Bank analysts estimate that the Palestinians' dependence on Israel importers increases the price of their produce by 20 to 25% compared to world prices (World Bank, 2006).

Time constraints on Palestinians

The confinement of Palestinians is not defined simply by the areas in which they can travel, but also by the temporal restrictions on their activities. All aspects of olive production are affected by this problem since separation imposes new routines and can sometimes prevent

4 The Palestinian Ministry of Agriculture estimates that these systems add 30% to the cost of export products for Palestinian companies compared to Israeli companies' costs for similar quantities (FAO, 2007: 22)

5 For example, some checkpoints used sniffer dogs for detecting traces of explosives in the merchandise. The Israeli authorities have now ordered that loads on palettes must be reduced from 1.8 m or more to 1.6 m to enable the dogs access to all the merchandise. This reduces the profit margin on exports because each truck now carries a smaller load and additional transport costs must be included in the bottle price.

access to olive orchards. Visitor's permits for the seam zone on the other side of the Separation Wall impose new schedules on Palestinian farmers: waking hours, working hours, number of days authorized for harvesting, etc. In particular, it is forbidden to spend the night on the land, even if the orchards are a long way from the farmer's home. Then, harvests are often disrupted by Israeli settlers (trees are uprooted, burned, cut down or stolen) and the number of incidence is increasing each year (OCHA, 2011a). Finally, total closure of the Territories' checkpoints imposed by the Israeli authorities for Jewish feast days or because of 'security threats' has a profound impact on both production and exports.

According to Véronique Bontemps (2012: 3), "*imposition of the principle of incertitude and of forced expectations is seen by Palestinians as alienating and humiliating, which reflects control of a dominant timetable on their dominated or 'expropriated' time*". Thus, time has become a new 'commodity' to which Palestinians do not have access (Peteet, 2008): they are forced to develop short-term strategies that allow them to continue to operate, but only as long as these strategies respect Israel's agenda[6] (Abu Sada, 2007). Collins (2008) has coined the term 'dromocracy', (from the Ancient Greek word *dromos* meaning 'race, speed'), for illustrating this phenomenon: control of time becomes a way of expressing the power and superiority of Israelis over Palestinians. Schedules for olive production are disrupted, but in a way that does not respect farmers' seasonal tasks or cultural and agricultural practices (Meneley, 2008b). 'Heterochrony', a biological term referring to the modification of the period and rapidity of an organism's development, can be used here to characterize the Israeli systems' ability to restrict the development of Palestinian social and economic structures.

Mount of Green Olives, a company that markets olives, offers a good example of how commercial practices have had to adjust to meet restrictions on exports, whether planned or unplanned. For example, the company's trucks are loaded with oil and taken to places close to the checkpoints the day before they cross the Wall. Commercial checkpoints are re-opened on Sundays once the Shabbat is over. The driver sleeps there until around 5.30 am so that he can be at the head of the queue at the security checkpoint and avoid heavy traffic caused by temporary or extended periods of closure. In addition, as there are a large number of Jewish religious feast days in April, September and October, the company re-organizes its activities in order to minimize disruptions to its annual export schedule and ensure that its commercial obligations are

[6] Caroline Abu Sada (2007) explains that PARC, an agricultural NGO, has changed its activities to include food safety programmes and has withdrawn from all its longer term projects such as the development of agricultural land.

met ahead of time. However, it is impossible to anticipate when restrictions will occur. For example, the company's storage and bottling factory is located at Awarta, near Nablus. In March 2011, the village was placed under a curfew for several days, following the assassination of a Israeli family living in a neighbouring colony. This total blockage was set up just when a consignment of olive oil was due to leave for Switzerland, France and Belgium. Controls were relaxed for one day to allow Mount of Green Olives to bring in glass bottles, but were strictly re-imposed the next day so that deliveries could not be made as promised to its European clients. This experience demonstrates the company's vulnerability to Israel's ability to cut off an entire village or territory for an unspecified period.

The delays imposed by Israeli control systems have had a two-fold impact on Palestinian olive oil production: on quality and on competitiveness. First, it is difficult for farmers to meet quality criteria for extra-virgin oil as the production process is extremely time-sensitive (e.g. speedy delivery to presses, daily checks on maturity of olives). Some agricultural produce, and particularly olive oil, is extremely vulnerable to being stored for long periods in inappropriate conditions (e.g. outdoors, lack of protection from sun, weather conditions). A third of Palestinian exporters have complained that, on at least one occasion, their merchandise has been spoilt during delivery (Haj Khalil, 2008). In addition, slowing down deliveries has reduced Palestinian olive producers' ability to react to market changes and thus their capacity to maintain long-term supplies that are required by the global market. They are unable to guarantee delivery dates imposed by an increasingly demanding consumer market. These problems create an additional handicap in the international market for olive oil where there is strong competition from traditional producing countries (Spain, Italy, North Africa) and emerging producing countries (China, Australia, Latin America, etc.).

Impossible routinization and self-confinement

Every day, Palestinians have to face a complex collection of regulations, inherited from previous systems introduced during the Ottoman Empire or the British mandate, and of laws adopted by Israel, Jordan and Palestine (World Bank, 2008). This multi-layered legal framework obviously opens up many opportunities for manœuvre but also accentuates uncertainties by generating contradictory dispositions. Not only is the absence of written and stable rules seen by Palestine as a form of control, but it is also used for that very purpose by Israel.[7] This arbitrary

---

7 *"Concerns about the trip are magnified by a sense of uncertainty. How many times have I heard people say that the real problem is not knowing the rules? At the beginning, I always used to say, 'There must be someone who decides what you can and*

system is thus used to restrict the life of individuals and the community by a series of precarious hesitancy and permanent readjustments. Palestinians thus are exposed to a high level of heteronomy (Bontemps, 2009), that is, the incapacity to establish their own laws and to behave in accordance with them. They are no longer masters of their social and economic future: on the contrary, they are at the mercy of an exogenous and erratic evolution of their situation from day to day.

In a recent novel on the 'country of checkpoints' that Palestine has become, Azmi Bishara (2004) emphasizes that the only routine possible is permanent change and this requires an endless reorganization of appointments, journeys backwards and forwards and leisure activities, depending on the latest information available. Thus, "paradoxically, the only certainty people have is uncertainty" (Romani, 2005: 259). The inability to organize journeys ahead of time upsets their ability to plan for their future in terms of space or of time. If, as Michel de Certeau (1990) has suggested, we accept that routine allows us to create a stable daily space, the absence or confiscation of routine creates a lack of guarantees for our most basic economic and social activities (Gazit, 2009: 89). Further, Riccardo Bocco (2005) refers to a multi-layered 'insecurization' of Palestinians' living conditions. This leads *in fine* to a form of self-confinement by Palestinians who renounce all willingness to set up a business or travel, in any circumstances, since the situation will probably deteriorate at any moment. For example, even though cars are now allowed through checkpoints, taxis continue to leave passengers on one side or the other of a checkpoint, because they fear that the road will be closed again (Handel, 2011: 270). Similarly, farmers may stay at home even after the end of a curfew, because they are afraid of meeting soldiers or settlers or that another curfew will be declared without warning (OXFAM, 2002: 18). Finally, farmers from Qalqilya and Tulkarem in northern West Bank remain in their home region and rarely travel to southern market towns, such as Hebron or Bethlehem. They are reticent to develop business opportunities that might fall apart at the first roadblock, since the further they have to travel to remote regions, the more difficulty they will have in returning (MAS, 2005).

Of course, it may be that, for most Palestinian farmers, the actual risk of confiscation or an attack by settlers is relatively low and that the impact on their activities is probably negligible. But, suggests Benoît Lallau (2008: 179), it is not enough that researchers and decision-

can't do!' Then I discovered that this void is a form of government" (Petti, 2008b). "*Vagueness and arbitrary situations produced by constant changes have been re-appropriated by the Israeli authorities. Rather than generating dysfunctions, they have become a way of controlling Palestinians so that they can no longer make travel plans*" (Parizot, 2010a: 9).

makers make certain assessments about the incidence or probability of a risk occurring in order to anticipate the impact on decisions made by individuals. Risks are perceived and amplified by the persons who are most exposed to them and have to act as if it represents a permanent threat.

The weight of this situation acts as a powerful brake on the economy, as can be seen in the activities of olive producers. In recent years, ministries and NGOs have introduced ambitious programmes which have resulted in a national 'revolution' in olive production by introducing three objectives: quantity, quality and marketing. Meeting these objectives requires very little investment, but a strong and long-term commitment by the farmers, according to Thomas Cazalis, Director of the Palestinian Farmers' Union (PFU)'s Olive Oil Programme:

> PFU's message is: with a small investment, we can create a revolution in the Palestinian olive oil production by agreeing to produce high quality goods which allow us to enter export markets. Of course it takes time. And time is one thing Palestinians do not have. It takes 5 to 6 years to change social attitudes and for the adoption of good practices to provide results. But Palestine is never sure of having 5 years of stability! Farmers are hesitant about making the investments that are necessary.[8]

Palestinians have to face a form of ontological instability and this does not encourage farmers to change their agricultural methods. The World Bank has made similar conclusions on the permanent state of uncertainly in the Palestinian Territories:

> Most problematic is the uncertainty caused by the closures. (...) If producers are not sure whether they will be able to produce sufficient olives, they will not be willing to invest and will therefore continue to reduce their financial risks by producing low quality bulk oil. (World Bank, 2006: 4)

Here I want to emphasize that ontological insecurity not only undermines daily activities, but it also changes the direction of all their business decisions. Other researchers (Petti, 2008; Handel, 2011; Havkin, 2008 and Parizot, 2010a, 2010b) have shown that Israeli control systems not only create specific methods for managing space and time, they also force Palestinians to initiate new commercial practices that take instability into account. Thus, some farmers may decide to concentrate on products that generate lower revenues, but are less likely to be affected by roadblocks or other difficulties for meeting delivery deadlines: for example, farmers no longer grow mandarins and grapes in the seam zone because they require high maintenance, preferring to grow other less demanding, but also less profitable products (UNOCHA, 2011).

8 Interview at Ramallah dated 25 April 2010.

Finally, many olive producers prefer to sell their olives immediately after pressing to private wholesalers at low prices, rather than seek more profitable orders by attending trade fairs at the end of the year.

The direct and indirect impact of Israeli control systems in the West Bank provides powerful arguments for limiting the development of the olive industry. Above all, they tend to destabilize Palestinians' commercial practices by depriving them of their ability to manage space, time and daily routines. Nevertheless, if we were to see these systems only as insurmountable and definitive obstacles, we would be, on the one hand, overestimating the systems put in place by the occupying power and, on the other, underestimating the ability of Palestinians to adapt to circumstances by side-stepping or exploiting separation mechanisms. In fact, our analysis of the micro-economy shows that the efficacy of control mechanisms is regularly challenged, while at the same time exposing the complexity of those same mechanisms. For Palestinians, this destabilization is now an integral element in planning their agricultural and commercial activities. They also seek to overcome the obstacles to their own mobility by finding easier solutions for transporting their merchandise. In the process, these adjustments allow them to reconfigure their relations with various other actors both inside and outside the West Bank.

Coming to terms with constraints: the emergence of an intermediary market

In order to exploit opportunities for negotiation and adaptation, some Palestinians have developed new socio-economic practices in response to the policy of separation. People who have strong business contacts, speak several languages, or have 'experience of crossings' manage to change their behaviour patterns and stand out from the crowd, even though they have to submit to the same systems of control. This has created new forms of unequal opportunities, a new social hierarchy, but *in fine* also serves to reinforce the existing level of dependency on Israel.

In fact, in order to continue their business and social activities despite the imposition of confinement, Palestinians must seek out other actors as 'go-betweens'. Go-betweens are either Israelis or people with a privileged position in the social hierarchy created by the matrix of Israeli controls.[9] This situation has also led to the emergence of 'a

9 There are a variety of status levels which determine how much mobility a person can have based whether they are Israeli Jew, Israeli Arab, or Palestinian permanent resident of Jerusalem, West Bank or Gaza, or have a foreign or a diplomatic passport, etc. (see Parizot, 2010a).

market of intermediaries' based not only on solidarity, but also on personal interests, cronyism and patronage. The result is a new form of business relations that reorganize power structures that bring these actors together, whether they are Palestinian, Israeli or other nationalities.

Palestinian actors

Since separation mechanisms increased during the early 2000s, there are now a wide variety of go-betweens who are helping Palestinians with production, trade or administrative procedures in the olive industry.

A first level of go-betweens comes into play when farmers are unable to gain access to their land. They will hand over responsibility for maintaining and harvesting their crops to tenant farmers who have a visitor's permit, in exchange for a share in the revenues. Next, because of fragmentation in production, the various phases of distribution are also fragmented. To ensure transportation of their produce, farmers will often hire Palestinians living in Jerusalem or others who have the appropriate permit[10] to make deliveries to markets in the Holy City or elsewhere in Israel. Sometimes, using their group of personal contacts can allow individuals to act as go-betweens without taking an active role. For example, Jalal,[11] a member of the Palestinian Olive Cooperative in Qarawa Beni Zeid, does not have a permit for entry into Israel and does not own any olive trees, but he earns two shekels for every kilo of the village's olive oil he sells to his contacts in Haifa and Nazareth. In a good year, he can deliver nearly 10 tons of oil in Israel by trading with contacts he made while working as a labourer in the 1990s. He takes advantage of informal trading networks he made earlier because other producers are prevented from establishing their own networks by the separation. Farmers are increasingly relying on their truck drivers to negotiate prices and contracts, in exchange for a commission on each transaction, because they are in regular contact with their clients. Thanks to this role as go-between, truck drivers are becoming powerful actors in the marketing cycle for produce (FAO, 2009). In the redistribution of roles, everyone can take on a new activity at any time as opportunities occur and as their personal status permits.[12]

---

[10] To obtain a permit, a person must make use of a long and costly network of various go-betweens.

[11] To protect interviewees, their names have been changed.

[12] Rema Hammami came to the same conclusion by studying operations at a checkpoint near Ramallah during the second Intifada. As cars were not authorized to cross, passengers had to cross on foot. Taxis and other transport companies lost clients and their business taken up by small dealers who used wheelbarrows and trolleys to transport people of the 2 kilometre section where cars were forbidden. One of the

Finally, at the administrative level, breaking up Palestinian national territory makes it difficult to ensure that public policies and the persons responsible for their implementation are consistent. For example, under the Oslo Accords, Areas B and C in the West Bank[13] were placed under Israel's partial or total control (respectively) in such a way that the Palestinian Authority has little say in their affairs. As a result, Palestinian agriculture is now managed through local NGOs, such as PARC and UAWC, which have a wider margin for manoeuvre. The NGOs are also responsible for managing joint programmes in the West Bank and the Gaza Strip, thanks to their many representatives in both regions, because the Palestinian Authority has difficulty imposing its policies, in part because Hamas has rejected its authority and in part because Israel issues only a limited number of permits to government representatives. A large proportion of funds from aid agencies and government is channelled through NGOs and is used to arrange for deliveries agriculture produce, payments of social benefits to farmers or networks for canalization. As such, they have become private go-betweens for government programmes and this has given them a powerful position in Palestinian social, economic and political affairs. According to Caroline Abu Sada (2007), most of the agricultural NGOs are backed by political parties which take advantage of this 'sympathy capital' for political purposes. For example, the current Minister for Agriculture, Ismail Deik, was previously Director of the Palestinian Agricultural Relief Committee (PARC), the largest agricultural NGO in the Palestinian Territories which is linked to the Palestinian People's Party.

Palestinian and Israeli actors

The reinforcement of control systems has increased Palestine's dependency on Israelis and thus required them to negotiate with the authorities and adapt their business strategies. The emergence of 'transfrontier' business partnerships, which create new forms of business practice, is now challenging and altering Israel's process of separation. Go-betweens may operate in the formal market, as mentioned above when describing the 'back-to-back' systems. Several thousands of

---

transporters said that "life is full of opportunities and when someone needs you, you can take the opportunity and earn a living". The author concluded: "the Checkpoint had offered men without rank an essential role in society's survival". In addition to a redistribution of roles, there are now new ways of redistributing profits (Hammami 2006).

13 Areas A are controlled by the Palestinian army and police; and Areas B by a coordination group of representatives of the Israeli army and the Palestinian police force. Areas B contain all the major Palestinian cities and towns. They represent about 40% of the West Bank in 2000. Areas C, which contain a number of Palestinian villages and hamlets as well as Israeli settlements, remain under Israeli control.

potential exporters are forced to unload their merchandise at one of six commercial checkpoints for the West Bank (Haj Khalil, 2008; Garb, 2011). After the goods have been checked, they are handed over to Israeli agents who are paid to put them on a second truck and ensure their delivery to commercial outlets in Israel and the rest of the world.

In addition, there are informal methods. Palestinians living in Israel play a key role in delivering oil from Palestinian mills to or via Israeli markets, thus bypassing Israeli and Palestinian accounting and traceability systems. The reopening of the Jalameh checkpoint for Palestinians in 2009 led to a period of strong growth in the olive industry in the northern West Bank's Jenine region. Samaritans, who are a minority population in this region, were hired as subcontractors for various activities. This little community, which lives in the Nablus hills in the West Bank and Holon in Israel, have the right to Israeli nationality and are therefore free to travel across the Green Line. Their ability to supply and deliver produce is very useful for West Bank inhabitants who are prevented from entering Israel (Bontemps, 2009; Doumani, 2004: 40). Finally, Jerusalem taxi drivers, whose identities and loads are not checked very carefully at border crossings, often make regular deliveries of small goods (cans of oil, pious objects in olive wood, etc.) from Bethlehem for sale in markets in the Holy City.

In addition, because export procedures and checkpoint passages are often uncertain, long and expensive, Palestinian producers may seek out and pay Israeli settlers to carry out certain services.[14] Settlers can take advantage of their status to make deliveries or as a cover for transporting merchandise discreetly and without complications between the various regions. Deliveries are usually improvised and arranged according to circumstances or, on the contrary, planned in advance by using highly organized networks of smugglers supervised by 'frontier entrepreneurs'. To a certain extent, these entrepreneurs become 'regulatory agents' who can authorize or facilitate the passage of individuals and goods (Natsheh and Parizot, 2011). They have a certain amount of power over those who make use of their services by obtaining or re-establishing opportunities for personal mobility that are so precious for people living in the Palestinian Territories.

Taking advantage of international actors

The Palestinians' need to adjust to local constraints also leads them to make use of foreigners as go-betweens. Because their movements are not restricted in the same way as for Palestinians, they can act as facili-

[14] See in particular Natsheh and Parizot (2011) for an illustration of these business partnerships.

tators and provide an international solution for restrictions on Palestinian movements.

Foreign priests and members of religious communities play a role in providing local or international transport for Palestinian produce. On the Mount of Olives in Jerusalem, a Spanish priest delivers the annual harvest of olives from the famous thousand-year-old trees in the Gardens of Gethsemane to an olive mill in Ramallah which has the reputation for being better and less expensive than mills in Jerusalem. Because of his dual status as priest and foreigner in the Holy Land, he can pass checkpoints more easily than Palestinian workers employed in Jerusalem. The vast network of religious communities can also act as transit agents or commercial outlets: for example, French parishes regularly sell Palestinian olive oil, which they order by the container load after their return from pilgrimages or via long-term cooperative activities.

On occasion, a go-between can operate in a number of different categories. In Taybeh, a Palestinian village near Jerusalem and east of the Separation Wall, the Coopération Catholique pour le Développement often sends French volunteers to work with its Catholic parish and its olive oil company, the Olive Branch Foundation. On their arrival, they are charged with making local deliveries of olive oil, incense and candles. With their western appearance and permits for driving cars with Israeli number plates, they can go through certain checkpoints or choose others where checks are less stringent. Their greater mobility makes it possible for produce to be delivered to market.

Furthermore, representatives of foreign donor agencies and diplomatic services regularly help the Palestinians by their contacts with Israeli authorities. First, they can obtain the necessary guarantees for establishing development projects which rely on approvals from the military administration. Donor agencies are also asked *ex-post* to intervene in the event of incidents with the Israeli forces relating to agricultural projects they have financed. For example, when water reservoirs were destroyed in an Area C near Hebron in July 2010, because they had been built without permits, the French agencies intervened:[15] use of go-betweens has taken on a political dimension. Finally, consulates and embassies arrange for Palestinian producers to receive a permit to come and sell their oil in Jerusalem's annual Christmas markets.

International solidarity also contributes. On the one hand, there is a growing number of foreigners who come to the country to help farmers

---

15 To understand the importance of these destructions and the involvement of donor agencies, it should be noted that Israeli authorities only issued 91 building permits between 2000 and 2007 in Area C which covers more than 60% of Palestinian Territories and most of Palestinian agricultural lands (McCarty, 2008)

throughout the year. Each year, several hundreds of persons participate in planting or harvesting programmes run by Palestinian and international organizations. Their involvement reduces the cost of harvesting and ensures that harvests are gathered in the best conditions possible, notably by discouraging attacks from Israeli settlers. Another form of cooperation is offered by fair trade programmes, which are often the only way that Palestinian oil can be exported. Fair trade also facilitates marketing campaigns that cannot be organized from Palestine "*because the difficulties faced by Palestinians wanting to travel overseas and by foreign buyers wanting to visit the West Bank*" (Benhayoun and Lazzeri, 2007: 112). Second, fair trade programmes provide partial or total funding for logistics for exports. This allows Palestinian producers to minimize the impact of Israeli systems in two ways: by transferring additional costs caused by the occupation to the foreign consumer; and, by reducing the distribution chain.

All these practices take into account, but also widen, the opportunities for using go-betweens. They reveal links of dependency, complementarity and solidarity amongst actors. While some involve financial accountability, others have a more moral dimension, what Marcel Mauss calls 'give-and-take'. Most of these practices lead to the emergence of new forms of 'bottom-up' partnerships based on power relations or personal interests, whether financial, social or political, for the actors involved.

Conclusion

Control systems have shaped Palestine's society and economy: directly, by restricting the movement of people and of goods; and indirectly, by discouraging entrepreneurial initiatives and by encouraging new methods of production and distribution. In spite of these control and separation systems, we can see that there are still many opportunities for adaptation and circumvention which the Palestinians are beginning to exploit. There are also opportunities which allow some actors to take advantage of their confinement and the vulnerability of others by acting as go-betweens from time to time. These forms of adaptation can also lead to an unexpected redistribution of roles, as anyone can take on the role of smuggler, informer, truck driver, etc. Individuals have learned to adapt their choices, ideas and careers to inconsistencies inherent in the political environment and the restrictions on mobility that are imposed on them.

In the Palestinian Territories, as elsewhere, the Wall and its checkpoints serve as an 'incentive to transgress'[16] which leads to a redistribution of roles amongst actors and moves towards more informal practices and sources of revenue. To survive, actors who find themselves under constraint must strive to re-establish opportunities of mobility for themselves or more importantly for their produce, even if this means they have to seek out a wide range of go-betweens. By attempting to come to terms with the constraints created by Israel's separation policy, they build partnerships that circumvent the control systems. Nevertheless, these readjustments do not directly challenge the symbols of domination but, on the contrary, reveal how solid and widespread they have become. It is not a question of interpreting them as acts of 'resistance' for this business sector (Abu Lughod, 1990). Even if the occupation confiscates the Palestinians' ability to manage their territory, it does not prevent them from reorienting, despite the precariousness of their situation, their socio-professional activities.

Readjustments to social and economic activities cannot be foreseen or planned. On the contrary, they must take place step by step and allow for improvisation according to new circumstances created by meetings, opportunities and constraints. These 'tactics' (De Certeau, 1990) are infinite with a wide range of methods and costs, depending on the actors involved and on time and space constraints, and they must be constantly readapted to keep step with an arbitrary and ever-changing environment. The actors, whom I have identified as 'go-betweens', are not necessarily professional smugglers: someone may be a smuggler one day and a client the next. Behind this apparent diversity, there is in fact a singular reality which is characterized by a similar and gradual attempt to re-appropriate one's private and public activities. As a result, a new relational form of business reinforces *in fine* Palestinians' dependence on each other, but also on Israelis – both Jews and Arabs – and on foreigners. This creates new domestic and international networks that cross social and spatial barriers established by the separation. Thus, we can see that the control systems have not succeeded in breaking off business activities between two closely intertwined areas, but rather have facilitated accommodations with the regulatory systems and a restructuration of power relations among the various actors in frontier areas.

16 "*Although very spectacular, the construction of walls does not interrupt mobility and it is even, like all challenges, a incentive for transgression, [...]. In these areas, the limitations set up by the frontier generate strategies for avoidance and resistance, force people to become inventive, and act as a real motivation for commercial and business activities. [...] For frontier economies, most forms of collaboration and cultural and social exchanges flourish through the informal market, by improvisation, and even clandestinity*" (Bennafla and Peraldi, 2008: 9).

References

Abu Lughod L. 1990. "The romance of resistance: tracing transformations of power through Bedouin women". *American Ethnologist* 17(1): 41-55.

Abu Sada C. 2007. "ONG palestiniennes et construction étatique. L'expérience de Palestinian Agricultural Relief Committees (PARC) dans les Territoires occupés palestiniens." 1983-2005. Beyrouth: Presses de l'IFPO: 254.

Abufarha N. 2008. "Land of symboles: cactus, poppies, orange and olive trees in Palestine." *Identities: Global Studies in Power and Culture.* 15(3): 342-238.

Arnon A. 2007. *Israeli Policy towards the Occupied Palestinian Territories: The Economic Dimension 1967-2007.* Université Ben Gourion du Neguev, July 2007: 35.

Benhayoun G. and Y. Lazzeri. 2007. *L'olivier en Méditerranée: du symbole à l'économie.* Paris: L'Harmattan: 140.

Bennafla K. and M. Peraldi. 2008. "Introduction. Frontières et logiques de passage: l'ordinaire des transgressions." *Cultures et conflits* No. 72: 160.

Bishara A. 2004. *Check point.* Translated from Arabic (Palestine) by Rashid Akel. Arles: Actes Sud, Collection Mondes Arabes.

Bocco R. 2005. "Les Palestiniens des territoires occupés face aux politiques israéliennes d'insécurisation". *Alternatives Sud*: 135-162.

Bontemps V. 2009 (not published). *Naplouse, le savon et la ville Patrimoine familial, travail ouvrier et mémoire au quotidien.* PhD Thesis. *Espaces, Cultures, Sociétés*, Université de Provence Aix-Marseille I.

Bontemps V. 2012. "Le temps de traverser le pont", in Temps et temporalités du monde arabe (XIXe-XXIe siècle). *Temporalités* No. 15, 2012.

Botiveau B. and N. Conte. 2005. "Après Gaza?. Introduction." *Études rurales* 1(173-174): 11-38.

Braverman I. 2009. *Planted Flags: Trees, Land and Law in Israel/Palestine.* Cambridge: Cambridge University Press: 272.

B'Tselem. 2007. *Ground to a Halt: Denial of Palestinians' Freedom of Movement in the West Bank.* Comprehensive Report: 177.

Collins J. 2008. "Dromocratic Palestine. The politics of time in Palestine". *Middle East Research and Information Project*, 248: 8-13.

Conseil Oléicole International. 2005. *L'oléiculture en Palestine.* E.108/Doc. No. 4. updated No. 32: 7.

De Certeau M. 1990. *L'invention du quotidien. 1 Arts de faire.* Paris. Gallimard: 347.

Doumani, B. 2004. « Scenes from daily life: the view from Nablus ». J*ournal of Palestine Studies* 34 (1), october: 37-50.

FAO. 2007. *Strengthening Resilience: Food Insecurity and Local Responses toFragmentation of the West Bank.* FAO, April 2007:80.

FAO. 2009. *The Impact of Closure and High Food Prices on Performance of Imported Staple Foods and Vegetable and Fruits Market in the Occupied Palestinian Territories.* December 2009: 220.

Farsakh L. 2005. *Palestinian Labour Migration to Israel: Labour, Land and Occupation. Routledge Political Economy of the Middle East and North Africa*. New York. Routledge: 286.

Fouet S. 2002. "L'étouffement des Palestiniens confinés dans des cantons". *Confluences Méditerranée* 4/2002 (No. 43): 23-32.

Garb Y. 2008. *A Survey of Freight Traffic between Israel and the Southern West Bank and the Tarqumia Crossing in Particular*. Ben Gurion University, Economic Cooperation Foundation: 32.

Garb Y. 2011. "Porosity, Fragmentation and Ignorance: Insights on Separation from a Study of Freight Traffic", in S. Latte Abdallah and C. Parizot, *À l'ombre du Mur: Israéliens et Palestiniens entre séparation et occupation*, Arles: Actes Sud: 125-152.

Gazit, N. 2009. « Social Agency, Spatial Practices, and Power: The Micro-Foundations of Fragmented Sovereignty in the Occupied Territories ». *International Journal of Politics, Culture, and Society* 22 (1) (mars 1): 83-103.

Grinberg L. 2011. "Discours économique et construction de la frontière dans l'espace israélo-palestinien depuis 1967", in S. Latte Abdallah and C. Parizot, *À l'ombre du Mur: Israéliens et Palestiniens entre séparation et occupation*, Arles: Actes Sud: 102-124.

Haj Khalil M. 2008. *Trade Security*. PSC Newsletter, Palestinian Shipper's Council. 1:6.

Hammami R. 2006. "De l'importance des voyous. L'éthique économique d'un check point." *Revue d'études palestiniennes* No. 98: 64-73.

Hanafi S. 2005. "Le projet colonial israélien: 'spatiocide' et 'biopolitique'". *Palestine: mémoire et perspectives* (Vol. XII 2005/1), Paris, Centre Tricontinental et Syllepse: 71-78.

Handel A. 2011. "Exclusionary surveillance and spatial uncertainty in the occupied Palestinian territories", in E. Zureik, D. Lyon and Y. Abu Laban (eds.), *Surveillance and Control in Israel/Palestine; Population, Territory and Power*. New York, NY: Routledge: 259-275.

Havkin, S. 2008. *La privatisation des checkpoints: esquisse d'une transformation d'un dispositif de pouvoir*. Masters 2 thesis in Philosophy, Université Paris 1 supervised by J. Benoist and E. Dorlin, September 2008. published on REASOPO, http://www.fasopo.org/reasopo/jr/mem_havkin.pdf.

Lallau B. 2008. "Les agriculteurs africains entre vulnérabilité et résilience. Pour une approche par les capabilités de la gestion des risques". *Revue française de socio-économie*, No. 1:177-198.

McCarty R. 2008. "Area C strikes fear into the heart of Palestinians as homes are destroyed". *The Guardian*, 15 April 2008, http://www.guardian.co.uk/world/2008/apr/15/israelandthepalestinians (Accessed 12.9.2012).

Mansouri W. 2007). "Les fonctionnaires de l'Autorité palestinienne et la suspension de l'aide internationale: notes d'enquête à Jenin (2006-2007)." *A contrario* 5(2): 174-183.

MAS (2005). *The Economics of Agriculture in the Tulkarem and Qalqilya Districts: Improving the Profitability of Farmers affected by the Separation Wall*. Ramallah: Palestine Economic Policy Research Institute: 98.

Meneley A. 2008a. "A tale of two itineraries: The production, consumption and circulation of Tuscan and Palestinian olive oil". Personal communication, 20 February 2009, *Colloquium Series 2008-2009*, Agrarian Studies, Department of Anthropology, Trent University.

Meneley A. 2008b. "Time in a bottle, The uneasy circulation of Palestinian olive oil. The politics of time in Palestine". *MERIP* No. 248: 22-27.

Natsheh B. and C. Parizot. 2011. "Du Kit-Kat au 4x4: la séparation sous l'angle du trafic de biens de consommation courante entre Israël et la Cisjordanie", in S. Latte Abdallah and C. Parizot, *À l'ombre du Mur: Israéliens et Palestiniens entre séparation et occupation*, Arles: Actes Sud: 153-184.

OCHA. 2010. *Access and Movement in the West Bank*. Annual Report. East Jerusalem, Occupied Palestinian territory, Office for the Coordination of Humanitarian Affairs.

OCHA. 2011a. *The Humanitarian Report*. December 2010/January 2011. East Jerusalem, Occupied Palestinian Territory, Office for the Coordination of Humanitarian Affairs.

OCHA. 2011b. *East Jerusalem, Key Humanitarian Concerns*. March 2011. East Jerusalem, Occupied Palestinian Territory, Office for the Coordination of Humanitarian Affairs.

OXFAM. 2002. *Forgotten Villages, Struggling to survive under closure in the West Bank*. OXFAM Briefing Papers, September 2002.

OXFAM. 2010. *The Road to Olive Farming – Challenges to Developing the Economy of Olive Oil in the West Bank*. OXFAM briefing paper, December 2010.

Parizot C. 2006. "Entrepreneurs without borders: Policies of closure and border economy between the Southern West Bank and Northern Negev, 2000-2005". *Europe and the Mediterranean Convergence, Conflicts and Crisis*. European Studies Centre, University of Oxford, RAMSES Working Paper.

Parizot C. 2009. "Après le mur: les représentations israéliennes de la séparation avec les Palestiniens". *Cultures & Conflits* No. 73: 53-72.

Parizot C. 2010a. "Séparation et régime de mobilité entre Israël et la Cisjordanie (2000-2009): Réglementations, pratiques et acteurs", in E. Benbassa (ed.), *Israël-Palestine, état des lieux* Paris: CNRS éditions: 129-146.

Parizot C. 2010b. "Temporalités et perceptions de la séparation entre Israéliens et Palestiniens", Bulletin du Centre de recherche français à Jérusalem [En ligne], 20 | 2009, placed online on 10 March 2010, http://bcrfj.revues.org/6291 (Accessed 12.9.2012).

Peteet J. 2008. "Stealing times. The politics of time in Palestine". *MERIP, Middle East Research and Information Project*. 248: 14-17.

Petti A. 2008. "Asymmetries", in A. Farquharson (ed.), *The Impossible Prison*. Nottingham: C.f.C. Art.

Petti A. 2008. "Asymmetries in Globalised Space: The Road Network in Palestine-Israël". *Henry Moore Institute – Online Papers & Proceedings*, http://www.henry-moore.org/hmi-journal/homepage/view-by-conference/still-life/alessandro-petti/page-2 (Accessed 12.9.2012).

Pirinoli C. 2005. "Effacer la Palestine pour construire Israël. Transformation du paysage et enracinement des identités nationales". *Études rurales. Palestine* 173-174(1-2): 67-85.

PMoA. 2008. *Monthly Report for Agricultural Losses during the Intifada*, 28/09/2000-31/11/2008. NovembER 2008. Ramallah, General Direcorate of Policy and Planning.

Romani V. 2005. "Quelques réflexions à propos des processus coercitifs dans les Territoires palestiniens". *Études rurales, EHESS*, No. 173-174: 50-70.

UNOCHA. 2010. *West Bank Movement and Access Update Report*. June 2010.

UNOCHA. 2012. *West Bank Movement and Access Update Report*. September 2012.

World Bank. 2006. *Brief Overview of the Olive and the Olive Oil Sector in the Palestinian Territories, Report.* World Bank, New York.

World Bank. 2007. *Movement and acces restrictions in the West Bank: uncertainty and inefficiency in the Palestinian economy*. World Bank, New York.

World Bank. 2008. *The Economic Effects of Restricted Access to Land in the West Bank. Comprehensive Report*. World Bank, New York.

PART III

BEYOND STATE BORDERS: MIGRANTS' MOBILITY IN EVERYDAY LIFE

Introduction

Lisa ANTEBY-YEMINI, Virginie BABY-COLLIN

Crossing international borders is only the first step taken by global migrants on their path, as they soon encounter internal boundaries in the country of migration which may be just as difficult to cross. Undocumented migrants and asylum-seekers are particularly vulnerable, as they have to deal with repressive control policies, attitudes and discourse of rejection in host societies, together with humanitarian forms of governance that are characteristic of today's global trends towards repression and protection in Western countries. Internal borders for migrants with different status, which vary over time and according to recent changes in public policies, are linked to different access to rights or denial of rights, which in turn affect the ways migrants live, move and project their future. Thus, the third section of this book enquires into everyday experiences of vulnerable migrants beyond state borders and in urban spaces where they struggle to settle. It aims at understanding not only the ways their illegitimacy or illegality is produced as a result of policies, discourses, attitudes or humanitarian interventions in the countries of destination, but also the different ways migrants handle their everyday life when crossing or bypassing those internal boundaries. Three contributions focus on different groups and spaces: undocumented migrant workers in Spain and the US (Baby-Collin); African asylum-seekers recently arrived in Israel (Anteby-Yemini); and women and independent child migrants working in the sex industry in several countries of the European Union (Mai).

A first set of questions running (through these articles pertains to the production of 'illegality' (De Genova, 2002) and the criminalization of migrants in various contexts, from the United States to Europe and the Middle East. In some cases, the state produces different status categories that make it easy for migrants to fall into the category of 'undocumented' or 'illegal alien', even if they were 'documented' before. The host population, media and civil society also construct images of migrants as vulnerable individuals or as criminals, either determined to have them included or demanding that they be excluded. Yet national and geographical borders do not always parallel social and political boundaries and migrants can sometimes cross, blur, or transform them; in addition it is the articulation between the symbolic and the social boundaries (class, gender, race, ethnicity) that is central (Lamont and Molnar, 2002) in the relation between migrants and the host country. It is these practices of governmental or informal construction of political and symbolic boundaries, as well as migrants' responses and practices of these borders, by contesting or avoiding the modes of control discussed in the first workshop, that interest us here. The gradual incorporation of borders involves a multiplication of status categories within the same spaces (undocumented migrants, asylum-seekers, temporary migrant workers, permanent residents, etc.) which are constantly redefined and transformed as different waves of migration follow one another over the long term and different policies evolve, sometimes in the same national space.

One result of the production of 'illegality'" is the fear that migrants develop of being arrested or deported and they will usually adapt their behaviour in order to avoid control and surveillance. These new mobilities influence the way migrants read and use urban, national and transnational space. There are therefore different territorial regimes – legitimate and illegitimate territories; territories in which people may be visible or remain invisible. These various dimensions may undergo changes in rules at different rhythms – coexisting in these contexts of embodied borders and status. These logics of invisibilisation or the geography of invisibility guided by fear of arrest and deportation affect migrants' modes of adaptation in their daily lives. Because of local variations in the application of laws and the discrepancy between national, state and municipal policies towards undocumented migrants, different territories emerge with their own regimes of mobility that migrants learn to appropriate; they develop ways to judge where it is best for them to live and circulate. This demonstrates how their daily mobility and their investment of space are affected by their specific status. In other words, circulating in the city without being arrested entails strategies not to be seen, such as taking taxis instead of public transportation, having parties in private houses instead of in public venues, or holding jobs requiring less visibility (e.g. domesticity) as a routine part of migrants' experience of 'illegality' (Baby-Collin). These spatial practices have also been observed in other global contexts of migration (Willen, 2007).

In addition, undocumented migrants' international mobility and local circulation is often constrained (Baby-Collin), even though re-emigration and circular migration is also observed among deported migrants who will migrate again, often with false documents, or continue their journey elsewhere (Mai). Furthermore, this production of 'illegality' and the increased tightening of internal and external borders reflect on migratory trajectories and social relations within migrant communities, as some migrants enjoy a status of temporary protection, for example, when others are being expelled (Anteby-Yemini).

Processes of invisibilization also lead to marginalization and exclusion yet, as the articles demonstrate, one also witnesses mechanisms of inclusion in modernity (Mai), in urban citizenship (Anteby-Yemini) and in social life in the host country (Baby-Collin). Therefore the daily experience of migration at the legal-political level, informed by laws, is sometimes quite different from the social and spatial practices of migrants. These articles thus show the ways in which governance operates, by comparing the priorities, discourses and practices of state and humanitarian interventions with those of the migrants in their daily life.

Finally, this section of the book, by looking at studies in the USA, Europe and the Middle East also seeks to take a critical approach to humanitarianism (Agier, 2011; Fassin, 2011; Ticktin, 2006). The question of whether migrants should only be seen as 'victims' is raised in these articles and requires us to explore alternative ways of representing the migratory experience where the agency of the migrants should be given an increasing role. Simultaneously, "humanitarianism has been transformed into a form of politics, functioning as a transnational system of governance tied to capital and labour even while purporting to be apolitical" (Ticktin, 2006: 33). Therefore, the consequences of humanitarianism as politics can include new bio-political practices that were unintended, such as undocumented migrants asking how to be infected by HIV to claim residency and papers in France (*ibid.*). If 'politics of compassion' have been studied to show how migrants with 'true victimhood' scripts obtain asylum or other humanitarian protection (Fassin, 2011), other ways of portraying migrants have been under-researched, such as critically analyzing sexual humanitarianism, for example: when adult and child migrants in the sex industry are 'othered' as vulnerable and trafficked victims, they are seen as individuals in need of protection, without taking into account their resilience and the fact that they can exert agency and empower themselves by selling sex (Mai), thus challenging new moral borders.

References

Agier M. 2011. *Managing the Undesirables: Refugee Camps and Humanitarian Government*, Cambridge: Polity Press.

De Genova N. 2002. "Migrant 'illegality' and deportability in everyday life". *Annual Review of Anthropology*, 31, 419-447.

Fassin D. 2011. "Policing borders, producing boundaries: the governmentality of immigration in dark times". *Annual Review of Anthropology*, 40, 213-226.

Lamont M. and V. Molnar. 2002. "The study of boundaries in the social sciences". *Annual Review of Sociology*, 28, 167-195.

Ticktin M. 2006. "Where ethics and politics meet: the violence of humanitarianism in France". *American Ethnologist*, 33:1, 33-49.

Willen S. (guest editor) 2007, special issue on "Exploring 'illegal' and 'irregular' migrants' lived experiences of law and State power". *International Migration*, 45:3.

Between Embodied Cosmopolitism and Sexual Humanitarianism

The Fractal Mobilities and Subjectivities of Migrants Working in the Sex Industry

Nick MAI

Social interventions relating to migrants who are considered to be vulnerable and, as such, entitled to social protection are embedded in contradictory migration and human rights regimes. They are characterized by difficulties caused by the expansion and pluralization of migrations (Smith and King, 2012), humanitarian considerations relating to universal human rights and social protection at the international level (Benhabib, 2004), and restriction of these rights for migrants at the national level on the basis of their nationality and citizenship (Lochak, 2007). Increasingly, people are moving for two reasons: to escape inequalities and conflicts due to neoliberal policies (Bauman, 2002), and to pursue personally and economically rewarding lives, even at the risk of exploitation and abuse (Lindquist, 2008). Migrants' complex experiences of vulnerability and resilience have been reductively simplified by the arrival of a de-politicized representation of the world, with its constitutive socio-cultural heterogeneity and growing economic inequality which is increasingly viewed in terms of polarization between a unified 'humanity' and individual victims in need of care (Ranciere, 2004; Agier, 2010: 39). Humanitarian rhetoric produces absolute victims: i.e. essentialized bare lives (Agamben, 1998) whose 'suffering bodies' (Fassin, 2005) become their constitutive ontological trait and are contained in specific spaces (i.e. permanent or temporary refugee camps, aid camps, shelters, etc.) through humanitarian projects. The process of verifying the credibility of the suffering of migrants in order to ensure that they are both helped and controlled is in fact as a biographical border between deportation and recognition. The ubiquity of these bordering mechanisms is part of the proliferation of dematerialized spatial and moral borders and of new surveillance mechanisms and control technologies in neoliberal times (Graham, 2006).

Sexual democracies

The late modern period is characterized by the hegemony of neoliberal ideologies that have transformed previous ideologies and practices of social solidarity and participation in social and economic life. Under pressure from progressive anti-racist and egalitarian movements and ideologies of the 1960s and of their fight against openly racial discourses, it has ceased to be socially and politically acceptable to exclude people from social protection and economic participation (Hewitt, 2005). Openly racial, gendered and class-based markers of identity have been subsumed into renewed and depoliticized categories of locality and mobility, as 'being from here' and 'going somewhere' became meaningful narratives of subjectification and belonging, demarcating the difference between 'making it' and not (Hickman *et al.*, 2012). As a consequence, 'being mobile' and 'being from somewhere' have become key late modern discourses for the reproduction, through the geography and physicality they designate, of symbolic and social forms of distinction and exclusion in non-openly racial and ethnic terms, 'through geography and physicality' (Skeggs, 2004: 50). In an interlinked and parallel way, gender and sexuality have become new de-politicized neoliberal narrative repertoires through which racial hierarchies of belonging and mobility are re-inscribed into national(ist) and other collective identities. The new public discourse on gender and sexuality in the West is consistent with the emergence of a new regime of neoliberal governmentality, sexual democracy (Fassin, 2010) that pathologizes and essentializes the gender and sexuality of minority ethnic and migrant groups in order to police Western moral and spatial boundaries. In this context, the focus on sexual violence experienced by (typically Muslim) constitutive 'migrant' others (Hall, 1996), such as women or sexual minorities, becomes a discourse of migrant control (Tiktin, 2008).

The pathologization and criminalization of specifically gendered and sexuality-related forms of migration in terms of trafficking and exploitation (Aradau, 2004) are part of emerging forms of 'neurotic' (Isin, 2004) sexual governance through the parallel production of moral panics (Weitzer, 2007) and related humanitarian interventions, or 'sexual humanitarianism'. These interventions seek to recreate the notion of a unified, West-centric and hierarchical humanity around essentialized and moralized understandings of secularism, gender and sexuality. Sexual humanitarianism operates by containing, through social interventions, the mobility of migrant groups that have been strategically essentialized and othered as 'pure' victims of sexual oppression and exploitation. Since migrants' nuanced understandings and experiences of ambivalence, vulnerability and resilience are obfuscated, only a minority receives appropriate support through sexual humanitarian social interventions. Such interventions tend to exacerbate, rather than reduce, migrants' vulnerability to exploitation as they are a vehicle for enforcing restrictive migration laws and lead to deportations. By separating the extreme victimhood of targeted others from a moralized, globalized and unified humanity, sexual humanitarianism also acts as a form of symbolic governance. It prevents citizens of the global North, which excludes underprivileged classes and marginalized places in the West, from reflecting on their own increasing exploitability, commodification and alienation in neoliberal times (Mai, 2013a).

Research focus, evidence and methodology

Containment of migration through sexual humanitarian interventions is a highly contested and under-researched area. This chapter will show the ways in which sexual humanitarian governance operates by comparing the priorities, discourses and practices of sexual humanitarian interventions with those of migrants they seek to rescue. It will compare that migration and social protection is experienced by two groups of migrants working in the sex industry who have been 'vulnerabilized', i.e. constructed and addressed as vulnerable by sexual humanitarian interventions:

– migrants working in the sex industry, identified as potential victims of trafficking; and

– independent child migrants, identified as potentially vulnerable because they have been separated from their family or are engaged in 'errant' forms of mobility.

This chapter aims to place the actual experiences of these two groups of 'vulnerabilized' migrants at the centre of current understanding of the nexus connecting migration, the sex industry and vulnerability. It will draw on the combined findings of two recent and interlinked bodies of research that take a common methodological approach based on a comparison of the more official and morally sustainable narratives emerging from semi-structured interviews and (participant) observations of the way these were embodied in everyday life. The first body of research, the narratives, focuses on migrants working or acting as agents in the global sex industry (Mai, 2013a) based on the 'Migrant Workers in the UK Sex Industry' project funded by the Economic and Social Research Council (ESRC). This research was carried out between 2007 and 2009 and collected 100 in-depth qualitative interviews with women, men and transgendered persons working at all levels in London's sex industry and from all the most relevant countries of origin. It also included 34 in-depth qualitative interviews with male agents in the sex trade carried out in Albania and Italy as part of a pilot project for the

International Organization for Migration (IOM) in Rome in order to establish a psycho-social profile of Albanian and Romanian traffickers.

The arguments developed in this article are also based on the findings of two research projects carried out in 2004-2006, one on the relationship between independent child and youth migration and the other on that between sex work and vulnerability to exploitation (Mai, 2009). The first studied the experience of minors and young adults from Eastern Europe (Romania and Albania) and North Africa (Algeria, Morocco and Tunisia) who migrated to the European Union (Belgium, France, Germany, Italy, the Netherlands and Spain) with the specific objective of basing their livelihood strategy on sex work. The study took place in 2004-2005 and covered 18 cities in all these countries. Its findings are based on the analysis of ethnographic observations and 82 interviews (50 with minors and young adults, 32 with key informants). The second study focused on the relationship between the migration of Romanian (male and female) minors to Italy (Rome), their involvement in illegal activities and on local initiatives of social intervention. This research was commissioned by Save the Children Italy and undertaken in Rome and in Romania (Bucharest and Craiova) in 2005-2006. Its findings were based on an analysis of ethnographic observations and 100 semi-structured interviews, 60 of which were with minors/young adults (40 male and 20 female) and 40 with key informants. In both studies, about 30% of respondents were declared minors at the time of the interview. Most of the minors were aged between 15 and 17.

Of trafficking and other governmental moral panics

Anti-trafficking moral panics and social interventions play a strategic role in the deployment of sexual humanitarianism. Estimates of the number of migrants working in the sex industry have often been manipulated by politicians and anti-sex work lobbyists into a moral panic situation (Weitzer, 2012). Contrary to existing evidence, the majority of migrant sex workers are presented as trafficked (Cusick *et al.*, 2009). The complex understandings and experiences of agency and vulnerability of migrant sex workers are obscured for political and fundraising purposes of different social actors (Carline, 2011). Undocumented migrants working in the global sex industry who are caught in the police's anti-trafficking operations will probably be 'voluntarily' deported unless they denounce their agents and/or partners as unambivalent exploiters and thus expose their families to dangerous retaliation at home (Surtees, 2007). This approach based on criminalization and law enforcement that has been adopted by anti-trafficking initiatives undermines their main objective because the number of convictions for trafficking remains relatively low and fails to identify and support the real victims of trafficking (Dottridge and Arocha, 2010). At the same time, victims of trafficking are given support if they can produce a standardized story of their exploitation regardless of the complexity of their ambivalent allegiances and resentments towards those who helped them and profited from them. These dynamics have a very negative impact on their sense of self (Giordano, 2008). As a consequence, the livelihoods of migrants working in the sex industry are disrupted and their entitlement to rights and social justice remain largely ignored and unattended (Zheng, 2010).

The following is a long excerpt from the interview with Alina, a 32 years old woman from Moldova. Alina was first trafficked to the UK in the late 1990s. She was deported but decided to return to London and, subsequently began to work in the sex industry again. In the extract that follows, she describes the 'rushed and unsophisticated' way she was detained and deported the first time, when she did not denounce her (female) trafficker. She also explains why she decided to go back to the sex industry and describes the constraints and opportunities she faced when she was 'rescued' by sexual social interventions.

> In 2000 there was a raid in one of the flats. I was arrested together with lots of girls. I spent 2 months in prison in Heathrow…it was a detention centre… It was very rushed and unsophisticated in those days. I spent two days in a cell, for what? The immigration people were tough; they did not even ask you if you were ok. I had met a nice Albanian customer in Soho, he felt sorry for me and he came to visit… I was deported with no money at all; I even had to ask my mother the money for the taxi from the airport. I felt guilty, ashamed. I had sold my body for nothing. I kept thinking 'what have I got left?' I felt guilty and worthless and I started thinking to go back… I also wanted to meet her again, the lady, and to see whether I could do something about it…
>
> – How did you manage to go back?
>
> With the help of my Albanian friend. I borrowed money from him and from the maid and bought a fake passport. The Albanian guy helped me with that… The Albanian guy helped me with that… He was nice. I paid him back by working in construction, but then I decided to go alone… It was impossible for me to keep working in construction. Very heavy physically and also the environment was difficult. I was the only woman amongst builders. So I got a newspaper and I responded to an advert for a job in a flat in Heathrow. It was something else. I was free to work when I wanted and with whom I wanted and to do what I was comfortable with, with customers. I used to keep half of the money and I was earning more than £700…on a bad day… I remember that was a good life!!! I could finally rest, help my family, go to nice places… even have holidays!!! There were other girls there, with pimps. Most of the Lithuanians and some of the Polish had Albanian pimps… But I was free. With my fake passport I could have a good

life. I got fed up in the end though. I was tired of being scared of the police… And I wanted to study...

– So, how did you get where you are now, what happened?

Well, what happened is that one day the immigration police came and I mentioned what had happened to me before. They had my fingerprints from the previous time…so I had to explain. As I mentioned the name of the trafficker they let me off the issue of having false papers and released me. I had to sign once a week and that was it. I now have got refugee status and I was recently given indefinite leave to remain This time the experience was much better, I mean with the police. I was explained everything and was supported.

– How were the support services?

They were good. People there are very kind and they know what they are doing, but there is no freedom at all. Some girls go actually back to work to be free. You cannot pass from having what you want when you want to £40 per week and only training, training, training… And also, working without a pimp is a completely different story and they should allow that, rather than telling people to stop selling sex.

Alina clearly marks the difference between being trafficked and a voluntary engagement in the sex industry. She explains how she was able to obtain legal migration status (despite her fake papers) which allowed her to enjoy rights and negotiate acceptable (for her) working conditions. Her interview shows that it is not sex work per se that defines exploitation, but whether or not the sex worker is able to enjoy the profits of his or her sex work. It also demonstrates that supportive relationships can exist between women and their clients and that these are often based mutual consent and respect. However, Alina's story is also an exceptionally positive one as, in the end, she obtains the status of 'victim'. Most women who denounce their traffickers are not recognized as victims and are therefore detained and deported during their processing, as was Amina when she was first arrested as part of sexual humanitarian anti-trafficking interventions.

In the global North, social intervention and research into the migration of minors are based on categories for legal and social protection such as 'unaccompanied minors/separated children', 'asylum-seeking children', 'child exploitation' and 'trafficking'. These categories fail to take into account the mix of vulnerability and resilience experienced by each migrant minor and young adult in its complexity and thus contribute to further exacerbating their vulnerability to exploitation. To ensure that social interventions do not exacerbate the social vulnerability of their target groups, practitioners must be able to understand the moral, legal and socio-economic criteria on the basis of which the child, and the system of relationships surrounding him, act. For instance, the way in which some Roma families living in Rome perceive intervention by social services bears witness to stark differences in interpreting the concept of 'protection of the child' and related principles and priorities. In the following extract, a father explains that a mother will take her child with her when she goes begging in order to avoid leaving the child alone while both parents are working. The intervention of the Italian authorities is seen as intrusive, partial and harmful, since it results in children being separated from their parents and thus ensuring that they will be alone!

> Look, in Rome... the situation is a mess... I was with my wife and my daughter... in the morning I left for work... and so did my wife.
>
> What work where you doing?
>
> I was working on a building site and my wife was begging. I was telling you...
>
> Yes, pardon me.
>
> That we couldn't leave the children in the camp, that if the police came they would take them to social services, because they may think they were abandoned... And so my wife took the girl with her... but not to make her beg as well, just to keep her with her. Well, the municipal police saw her, then what happened is that they took the girl... Because they said she was using her to get more money... But that's not true at all, it's just that we couldn't leave her at home, nor take her with us to work...
>
> But did this also happen to other families?
>
> Even worse, there were many families in the camp who were sent away, and the children remained there... they were accompanied before that... it's the Italian government that made them become unaccompanied, that's what's happened...they had a family before!
>
> (Male, Romanian, 35, Craiova)

Increasingly, studies of prostitution refer to the 'sexual exploitation of children' when the person selling sex is younger than 18, since the term 'child prostitution' implies that the activity can be voluntary. The very possibility that a person under 18 might choose to sell sex is unacceptable for most researchers and practitioners (O'Connell Davidson, 2005: 23), which means that child prostitution is seen as coinciding with sexual exploitation. While it is important to recognize children's vulnerability to exploitation when selling sex, it is equally important to insist on the possibility that this is in fact a voluntary act of child prostitution. Acknowledging that minors can decide to sell sex makes it easier to understand their needs and desires as a whole and not just from the perspective of humanitarian and North-centric (Burman, 2008: 106, 188) conceptualizations of childhood and associated stereotypical notions of victimhood. Experience in the field demonstrates that minors voluntarily decide to sell sex rather than accept working conditions that are perceived as being more disadvantageous. The following excerpt from an interview with Jonutz, a 17 year old Romanian selling sex in Rome, offers a clear example of this choice.

How do you live? What kind of work do you do?

I live as best I can, I steal from shops then I sell what I steal at half price.

Do your friends do the same work too?

Yes, they do, all four. We also hang out in Piazza della Repubblica, that's where we met.

Do your parents know what kind of work you do?

No, they believe I'm working at a petrol station. That's what I told them because my parents wouldn't understand what I am doing, they're a bit old-fashioned...

Why do you say that your parents are old-fashioned?

Because they have always worked in the country and they haven't travelled to see the world, then if they knew that I have sex with men... my mother would die.

And are you satisfied with what you are doing in Italy?

Yes, in any case I don't work hard and I always have money in my pocket. I even managed to send my father 2000 Euros, so he could buy a tractor.

And so you want to return to Romania? Since you are sending money to your father and he is buying farming machinery, do you plan to work the land together with your father?

No, but he kept nagging that I wasn't helping him, so I sent money to make him happy. I'll never become a country farmer, since I saw the West I couldn't even live in the little village I come from.

(Male, Romanian, 17, Rome)

This last excerpt shows that many migrant male and female minors and young adults can exert agency and empower themselves by selling sex. Nevertheless, this possibility is increasingly denied by governmental and non-governmental guidelines and policies and by the rhetoric of victimhood on which these are based. Unfortunately, this humanitarian, infantilizing and North-centric approach to adolescence and the global South threatens to ignore important differences in the way the passage to adulthood occurs both within and outside the North. In addition, by underestimating the different degrees of agency and sexual awareness in the lives of minors and young adults selling sex, sexual humanitarian 'protectionist' narratives often contribute to their marginalization and stigmatization. As a result, many migrant minors and young adults turn away from opportunities offered by social protection and support organizations, which they recognize as attempts at infantilization, and prefer to find solutions for their needs in marginalized environments and illegal or stigmatized livelihood strategies, including sex work.

Between embodied cosmopolitism and sexual humanitarianism

The interplay between globalization, individualization and feminization of survival (Sassen, 2002) coincides with the renegotiation of hierarchies structuring local economies, collective and social imaginaries and related models of personhood and authority. Global media and returning migrants disseminate new embodied scripts of individualization that are embedded in material environments, lifestyles and models of personhood. They offer new opportunities for self-realization and also increase the burden of self-reflexivity and self-management for many minors and young adults worldwide (Elliott and Lemert, 2006: 172-4). All global societies are involved in these dynamics which, in the South, coincide with what Jiménez Álvarez (2004) calls the 'minorization of poverty'. As a result, more and more minors and young adults are now responsible for the economic viability of their households and have to assume an adult role at a very young age. Contemporary society is marked by the dissemination of a globalized and individualized 'moral economy in which economic success is a necessity' (Lindquist, 2008: 231). As part of this process, the imperative of individual economic affirmation apparently renders established authorities, gender roles and the associated livelihoods redundant. These wider dynamics emerge clearly from the body of research that forms the basis for this article. For most interviewees, migration is a way to mediate between their established role as sons and daughters providing for their families and the desire to experience a 'more modern' – i.e. individualized and hedonistic – lifestyle (Mai, 2004a).

The convergence of greater social inequality, higher expectations of individual economic success, and faster processes of individualization has produced forms of vulnerability and resilience that are shared by migrants and non-migrant citizens throughout the world. At an historic moment where the social experience of being a 'child', 'adult', 'man', 'woman', and non-heteronormative subject (gay, lesbian, queer, transgender, etc.) is being transformed in countries of origin and of destination, the dreams and realities of an individual's decision to migrate are central to the elaboration of new subjectivities, roles and positions. They emerge at the intersection of new globalized articulations of ethnicity, class, gender and sexuality and coincide with a range of different and complex socio-economic and cultural experiences of vulnerability and resilience. From this perspective, those leaving should not be considered merely as vulnerable 'victims' of heightened social transformations, but as 'social innovators' (Lăzăroiu and Alexandru, 2003: 22) encountering difficult challenges when their gender, sexual and migratory orientations emerge and confront different and transforming models of personhood, gender roles, economic priorities and moralities.

In this context, 'making money' emerges as a key discourse and a priority for independent young migrants. In addition to its obvious economic meaning, 'making money' is the 'shorthand' used by young people to justify their desire to achieve psychological autonomy and upward social mobility in a context characterized by poverty and the need to survive. It is part of the complex interplay of survival, individualization, aspiration to hedonistic lifestyles, adolescent attachment/detachment from the family, and transformation of models of authority that shapes the contradictory and complex needs of migrant young people. From this perspective, migration can be seen as a response to the necessity of negotiating detachment/autonomy from parents, contributing to the economy of the family, exploring new places, having fun, having sex, earning enough money to marry, showing off to the peer group/neighbours and looking for alternative figures of affective attachment: in a word, to grow up. It is within the interaction between this complexity and individual life trajectories of each single young migrant that aspects of vulnerability and/or resilience of their migratory selves can be identified.

By engaging in the global sex industry, young migrant men and women challenge the boundaries and hierarchies limiting the expression of their social mobilities and individualized identities at home. In the process, they experience and produce globalized 'new sexualities' (Hoefinger, 2011) at a time when new and old models of personhoods, techniques of the self, materialities and sex/gender regimes come into play (Mai, 2004b). Young male and female migrants, whether or not they are sex workers and regardless of their age, can afford, morally and economically, 'foreign' and cosmopolitan individualized and embodied lifestyles which are 'fractally queer' in relation to established sexual/gender roles 'at home' (*idem*). The concept of fractal is inherently contradictory as it seeks to capture the ambivalence between irregularity (queerness) and recurrence (normativity) and the consistency between the fragment (margin) and the whole (mainstream) (Mandelbrot, 1982). The geometrical metaphor of the fractal was recently introduced in social theory in order to emphasize that the proliferation of 'irregular' socio-economically marginalized spaces is consistent with the globalization of neoliberal post-industrialism (Soja, 2000). More recently, the fractal concept was used to stress that some of these globalized interstitial spaces are also spaces of creativity and experimentation, particularly in terms of negotiation of gender and sexuality in post-communist societies (Roman, 2007, 128-29). It is this last interpretation of the fractal social space as one of both irregular queerness and normative recurrence that I use in my work in general. The global sex industry and migration can be experienced as fractally queer diaspora spaces (Mai, 2012b) in that they allow young women and men, including minors, to engage in individualized practices and lifestyles, that are viewed as both irregularly queer and recurrently normative because they challenge and reproduce established hetero-normative life trajectories and patriarchal gender/sexual roles without taking full public responsibility for their actions 'at home' (Gal and Kligman, 2000).

In many accounts, the sex industry emerges as intrinsically and ambivalently queer, compared to hetero- and homo-normative 'straight' lives and jobs. Because they live a double life between their 'straight' lives and jobs and the sex industry and their earnings from selling sex, many migrants develop a sense of spatial and temporal separation that allows them to 'escape' from normative everyday life (Day, 2007: 87). Many are able to negotiate different aspects of themselves at home and abroad, as shown in the following excerpt from the interview with Cindy, a 'straight' 27 year-old Lithuanian woman working in a walk-up in Central London:

> I like it as well like this, that it is so separate. I have my own space and my own money. Here I am a bad girl ...while at home I am a virgin, so to speak. It is true that I am quite traditional though. I mean... I would never go to clubs and bring men home the way girls do here... At the same time I am traditional my own way... because look at what I do! (laugh) (...) The truth is that I am myself in both places, both at home and here... It's just different parts... At the same time it is quite difficult to keep up and live with all the lying... My family does not know anything, of course! And only a couple of friends know...

Cindy's 'transgressive' and 'traditional' positions express the fractal ambivalence towards normativity and queerness that defines the global sex industry as a whole. By oscillating between 'home' and working in a brothel, she enjoys conforming to and transgressing normatively defined parts of herself, i.e. the 'bad girl' and the 'virgin', through embodied practices. At the same time, she decides fractally not to assume a unified social and personal identity in gender, sexual or professional (i.e. sex worker) terms by keeping the two aspects separate. The fractal ambivalence between challenging and reproducing normativity is a constitutive element of the sex industry, which is defined by its unique combination of queerness, marginalization and centrality within society.

The migrants I interviewed in the context of my research were challenging the restrictions imposed on the transformation of their gender, sexual, socio-economic and cultural orientations through tacit (Decena, 2011) bodily practices, including selling sex and migration, rather than through discursive practices. In order to underline the relevance of class and social capital in the adoption of embodied, rather than discursive, practices in challenging existing norms and restrictions, I introduce the concept of 'boditarian cosmopolitism' (Mai, 2011). Just like 'proletarian', the term 'boditarian' aims to capture the complex interplay between class-based marginalization, resistance, vulnerability and resilience. Most importantly, the term emphasizes the relevance of the bodily, over the self-reflexive and verbalized, dimension in people's experiences of self. By circulating between normative worlds and established patterns of integration, migrants engage in new forms of 'cosmopolitism from below' coinciding with the formation of new hybridized practices and identities (Tarrius, 2000) and the emergence of new subaltern cosmopolitan political practices (Bonaventura de Sousa, 2005). Young migrants working in the global sex industry are potentially 'vernacular cosmopolitans' (Werbner, 2006), who can only afford (economically and morally) to enjoy globalized neoliberal lifestyles by migrating abroad.

Young migrants' need to 'make money' transcends the economic dimension and reflects a moral and risk economy where the social stigma attached to returning home 'empty-handed' is far greater than that associated to the danger of being exploited in the process (Lindquist, 2008: 227). During this process, migrating in order to 'make money' at all costs, including by selling sex and engaging in illegal activities, can become a morally acceptable way to negotiate an adult role through detachment from one's family of origin and the establishment of one's own family and livelihood. Interviews with Milena, a 26 year-old woman selling and managing sex in Italy, and Adrian, her fellow-Romanian 25 year-old agent/boyfriend, demonstrate the fractal tension between emerging individualization and the resilience of collective social formations and affiliations at work:

> We both come from a bad place. We always had to do things that we did not want to do...now I feel I don't have to lead that shit life anymore. Where I come from is a shithole, there is nothing to do and the mentality is very old, if you want to have a good time you have to become somebody's whore, otherwise they get you to marry and create a family and then you are old as

> they are. I thought that the situation was better here (Italy) than over there and it was true. Everybody was saying the same thing. One way or another everyone who came here made money and so when I met him...and I understood he was making money, I decided to come here too (Alina).

> To live in Romania nowadays…you do not really need money to live, but to prove to others that you are better than them…sometimes you live like shit but outside people think you are ok… (...) Look at me, I am full of gold and so is my girlfriend. Look at Romanian women or girls; they are always full of gold because they need to prove to somebody that they are doing well. (...) The problem is this always having to prove something, to people and also at home, to your family, they expect everything from you and when you go out you need to be able to show that you can afford fashionable things, otherwise you are not worth a fucking thing (Adrian).

Milena and Adrian highlight the shared socio-economic and cultural conditions of men and women who decide to migrate in order to 'make money' as workers and agents in the global sex industry. In the process, established and gendered models of personhood are both reinforced and challenged. Young people's migratory and sexual orientations straddle the aspiration to an individualized and more hedonistic lifestyle, the stigmatization of women wanting to have 'a good time' and the performance of standardized deeds for economic success and privileges 'at home'.

Contrary to the prevailing understanding of the nexus between migration and the sex industry as marked by the prevalence of exploitation – which is reduced conceptually to the ruthless exploitation of female victims by male villains – only a minority of migrant women working in the sex industry see themselves as exploited, let alone trafficked (Platt *et al.*, 2011). In the research analysed for this chapter, the boundaries between advantage and disadvantage and between love and exploitation are complex and ambivalent. Migrants' perception and experience of enjoying fair working conditions or being subject to exploitative behaviour are linked to the perception and experience of their primary economic/professional interests – being able to work and be paid and treated appropriately in the sex industry – that are being undermined. However, the definition of 'appropriately' is also embedded in gendered cultural constructions, roles and practices. The perceived transgression of agreed negotiations and the denial of their very entitlement to negotiate their best interest are at the root of women's experiences and their understandings of violence and abuse, not sex work per se nor being managed by men per se. In the context of independent and male child and youth migration, we came across a few cases where young men managed other young(er) men. These circumstances were not understood as 'exploitation' but were morally justified in peer talk by references to the need for older young men to 'tutor' younger and inexperienced 'boys'. In the second research project on independent and child migration, a few of the male minors and young adults who were selling sex reported that they had been mugged and threatened by older adults and mentioned cases of 'exploitation' of minors involved in stealing and begging. In contrast to these experiences, many interviewees stressed that the vast majority of minors and young adults sold sex voluntarily, as sex work came to be perceived 'a good job', compared to other irregular jobs available and to the precarization of their livelihoods at home.

> Do you know whether there are small kids who work as rent boys here?
>
> Yes there are, 13 or 14 years old... and they have customers and work too, we all come from the same town and we know one another. We know that this is a good job. What do I do in Romania, must I wait for someone who asks me to throw out a table or cabinet from his house for him for 50,000 Lei, what do I do with 50,000 Lei (1.5 Euro)?
>
> (Male, Romanian, 19, Rome)

Restrictive migration policies and the lack of legal migration documents were often mentioned as the main causes behind women's entanglement in exploitative situations. According to Mariska, a 20 year-old woman from the Czech Republic working in the UK,

> When I talk with my work friends I know that the main thing is the documents. After boyfriends becoming 'pimps' I think documents are the main problem for the working girls. The situation can become better by helping people from outside the EU to get more easily working visas or papers.

Mariska's observations are very important for the argument here because instances of exploitation tend to occur almost exclusively with migrants who are undocumented. The implementation of 'protectionist' social interventions and restrictive migration policies can exacerbate minor and young migrants' vulnerability to exploitation. For instance, the Italian government introduced new legislation in the early 2000s that changed the criteria for the regularization of minors: minors who arrived in Italy before the age of 15 and who participated in a social reintegration project for at least two years were now eligible. This provision was meant to help unaccompanied migrant children by providing them with greater opportunities of integration when they become adult at 18. However, this law did not correspond to their priorities and needs as it forced even younger children to take responsibility for the survival of their families by migrating and remitting home. Immigration status is by far the most important factor in restricting minor and adult migrant sex workers' ability to exercise their rights in their professional and private lives. By working in the sex industry, many young migrants keep a low profile and avoid visibility: this protects them from being controlled and deported when working in other sectors of employment. Calling for the police in dangerous situations is an option available to documented migrants and undocumented migrants who are not afraid of being exposed as sex workers at home, since police operations are often associated with the very real possibility of deportation. In this context, male partners, even when they 'become a pimp' are sometimes viewed as an ambivalent source of both protection and oppression because they both facilitate and control women's travel to the West despite existing restrictions on their entry into the labour market in most EU countries. At the same time, they protect women from the stigma associated with sex work, by assuming the role of 'official partner' in the eyes of the community of origin, while controlling them through threats to denounce them as sex workers to the same home community.

Conclusion

Sexual humanitarian discourses equate child prostitution with child exploitation. They strategically construct all migrant sex workers as trafficked and their partners and agents as exploiting pimps. In the process, the 'vernacular' and 'boditarian' plurality of interpersonal dynamics, subjective investments and migration trajectories as evidenced by scientific research is ignored and obfuscated. At the same time, North-centric international norms relating to trafficking in persons and to children and women's human rights have become vernacularized at the national level as anti-prostitution and anti-migration policies (Cheng, 2011). Gendered, ethnicized, class-based and age-based hierarchies of entitlement to social and spatial mobility are enforced through the rhetoric and practices of humanitarian interventions. As this polarized and hierarchical world does not match the needs and priorities of the people involved, sexual humanitarian initiatives are perceived by migrant minors and young adults working in the sex industry as a threat, rather than as an opportunity. Certification of the credibility of migrant's victimhood stories, which is embedded in the proliferation of sexual humanitarian biographical borders, acts as a powerful mechanism for migration control. Only a very small minority of 'true' victims of exploitation are allowed to stay, all other biographies being systematically investigated and their subjects deported. Important economic resources are wasted, as the emergence of rigid biographical borders impedes the recognition of victims and their access to fundamental rights and social protection. In the process, the vulnerability of migrant minors and young adult women selling sex is exacerbated as they become more dependent on friends and partners in order to keep working and earning, while being treated, appropriately in the global sex industry. These dynamics highlight a key observation: concepts and experiences of exploitation and abuse must be analyzed with reference to the cultural and social realities in which interviewees live and evolve. What defines exploitation in the eyes of migrants is not involvement in illegal or stigmatized activities per se, but their inability to benefit from their involvement in terms of autonomy and economic profit. Stealing, selling sex and begging can be perceived as morally acceptable livelihood strategies, as long as they generate enough money to maintain their migratory route towards autonomy and economic self-realization. In the context of globalized and neoliberal moral economies that encompass the migratory journeys of migrants working in the global sex industry, 'making money' is a key discourse and existential priority, legitimizing the moral viability of their boditarian cosmopolitan mobilities and of their 'fractal' identities and practices. By denying the fractal queerness and cosmopolitanism of migrants working in the global sex industry, sexual humanitarian interventions criminalize their entitlements to social and spatial mobility. The associated moral panics over the trafficking and sexual exploitation of migrants prevent citizens of 'advanced' neoliberal capitalist societies from reflecting on their own increasing exploitability, commodification and alienation in neoliberal times.

References

Agamben G. 1998. *Homo Sacer: Sovereign Power and Bare Life*. Stanford: Stanford University Press.

Agier, M. 2010. "Humanity as an Identity and Its Political Effects (A Note on Camps and Humanitarian Government)". *Humanity: An International Journal of Human Rights, Humanitarianism, and Development*, 1(1): 29-45.

Ahmed S. 2006. "Orientations: Towards a Queer Phenomenology". *GLQ*, 12, 4: 543-574.

Aradau C. 2004. "The Perverse Politics of Four Letter Words: Risk and Pity in the Securitisation of Human Trafficking". *Millenium: Journal of International Studies*, 33(2): 251-277.

Bauman Z. 2002. *Wasted Lives: Modernity and its Outcasts*. Cambridge: Polity Press.

Benhabib S. 2004. *The Rights of Others: Aliens, Residents and Citizens*. Cambridge: Cambridge University Press.

Bonaventura de Sousa S. 2005. *Toward a New Legal Common Sense: Law, Globalisation and Emancipation*. Cambridge: Cambridge University Press.

Burman E. 2008. *Developments: Child, Image, Nation*. London: Routledge.

Carline A. 2011. "Criminal justice, extreme pornography and prostitution: Protecting women or promoting morality?" *Sexualities* 14(3): 312-333.

Cheng S. 2011. "The Paradox of Vernacularization: Women's Human Rights and the Gendering of Nationhood". *Anthropological Quarterly*, 84: 475-505.

Cusick L., H. Kinnell, B. Brooks-Gordon and R. Campbell. 2009. "Wild guesses and conflated meanings? Estimating the size of the sex worker population in Britain". *Critical Social Policy* 29(4): 703-719.

Day S. 2007. *On the Game*. London: Pluto.

Decena C. 2011. *Tacit Subjects: Belonging and Same Sex Desire among Dominican Immigrant Men*. Durham and London, Duke University Press.

Dottridge M. and L. Arocha. 2010. *Wrong Kind of Victim – One year on, an analysis of UK measures to protect trafficked persons*. London: Anti-Trafficking Monitoring Group. Available online: http://www.antislavery.org/includes/documents/cm_docs/2010/a/1_atmg_report_for_web.pdf.

Elliott A. and C. Lemert. 2006. *The New Individualism*. London: Routledge.

Fassin D. 2005. "Compassion and repression: the moral economy of immigration policies in France". *Cultural Anthropology*, 20(3): 362-387.

Fassin E. 2010. "National identities and transnational intimacies: sexual democracy and the politics of immigration in Europe". *Public Culture* 22(3): 507-529.

Gal S. and G. Kligman G. 2000. *The Politics of Gender after Socialism: A Comparative-Historical Essay*. Princeton: Princeton University Press.

Giordano C. 2008. "Practices of translations and the making of migrant subjectivities in contemporary Italy". *American Ethnologist*, 35(4): 588-606.

Graham S. 2011. *Cities Under Siege: The New Military Urbanism*, London: Verso.

Hall S. 1996. "Introduction: who needs identity?" in S. Hall and P. Du Gay (eds.), *Questions of Cultural Identity*. London: Routledge, 1-17.

Hewitt R. 2005. *White Backlash and the Politics of Multiculturalism*. Cambridge: Cambridge University Press.

Hickman, M., N. Mai and H. Crowley 2012. *Migration and Social Cohesion in the UK*, Basingstoke: Palgrave Macmillan.

Hoefinger H. 2011. "'Professional Girlfriends': an ethnography of sexuality, solidarity and subculture in Cambodia". *Cultural Studies*, 25(2): 244-266.

Isin E.F. 2004. "The Neurotic Citizen". *Citizenship Studies*, 8(3): 217-235.

Jiménez Álvarez M. 2004. *Análisis transnacional de los procesos migratorios de los menores de origen marroquí*, IV Congreso de la Inmigración. Girona, Spain.

Lăzăroiu S. and M. Alexandru. 2003. *Who is the Next Victim? Vulnerability of Young Romanian Women to Trafficking in Human Beings*. Bucharest: IOM.

Lindquist J. 2008. "Of maids and prostitutes: Indonesian female migrants in the new Asian hinterland", in M. DelVecchio, S.T. Hyde, S. Pinto, and B.J. Good (eds.) *Postcolonial Disorders*. London: University of California Press, 218-237.

Lochak D. 2007. *Face aux migrants: état de droit ou état de siège?* Paris: Éditions Textuel.

Mai N. 2004a. "'Looking for a more modern life…': the role of Italian television in the Albanian migration to Italy". *Westminster Papers in Communication and Culture*, 1(1), 3-22. Available online: http://www.wmin.ac.uk/mad/pdf/Mai.pdf.

Mai N. 2004b. "Albanian masculinities, sex work and migration: homosexuality, AIDS and other moral threats", in Worton M. and Wilson-Tagoe N. (eds.) *National Healths: Gender, Sexuality and Health in a Cross-cultural Context*. London: UCL Press: 45-58.

Mai N. 2009. *Migrant Workers in the UK sex industry* – Final Policy-Relevant Report, ESRC final project report. Available online: http://www.londonmet.ac.uk/research-units/iset/projects/esrc-migrant-workers.cfm.

Mai N. 2011. "Tampering with the sex of "angels": migrant male minors and young adults selling sex in the EU". *Journal of Ethnic and Migration Studies*, 37(8): 1237-1252.

Mai N. 2012. "The Fractal Queerness of Non-Heteronormative Migrants Working in the UK Sex Industry", *Sexualities*, 15(5-6): 570-585.

Mai N. 2013. "Embodied Cosmopolitanisms: the Subjective Mobility of Migrants Working in the Global Sex Industry", *Gender, Place and Culture*, 20(1): 107-124.

Mandelbrot B.B. 1982. *The Fractal Geometry of Nature*. New York: Freeman.

O'Connel Davidson J. 2005. *Children in the Global Sex Trade*, Cambridge: Polity Press.

Platt L., P. Grenfell, C. Bonell, S. Creighton, K. Wellings, J. Parry and T. Rhodes. 2011. "Risk of sexually transmitted infections and violence among indoor-working female sex workers in London: the effect of migration from Eastern Europe". *Sexually Transmitted Infections*, 87(5): 377-84.

Rancière J. 2004. "Who is the subject of the rights of man?" *South Atlantic Quarterly*, *103*(2/3): 297-310.

Roman D. 2007. *Fragmented Identities. Popular Culture, Sex and Everyday Life in Postcommunist Romania*. Lanham: Lexington Books.

Sassen S. 2002. *Counter-Geographies of Globalization. Feminization of Survival*. London: Zed Books, 89-104.

Skeggs B. 2004. *Class, Self, Culture*. London: Routledge.

Smith D. and R. King. 2012. "Editorial introduction: re-making migration theory". *Population Space Place*, 18: 127-133.

Soja E. 2000. *Postmetropolis, Critical Studies of Cities and Regions*. Oxford: Blackwell.

Surtees R. 2007. *Listening to Victims: Experiences of Identification, Return and Assistance in South-Eastern Europe*. Vienna: ICMPD.

Tarrius A. 2000. *Les Nouveaux Cosmopolitismes. Mobilités, Identités, Territoires.* Paris: Éditions de l'Aube.

Tiktin M. 2008. "Sexual violence as the language of border control: where French feminist and anti-immigrant rhetoric meet". *Signs: Journal of Women and Culture in Society* 33 (4): 863-889.

Weitzer R. 2007. "The social construction of sex trafficking: ideology and institutionalization of a moral crusade". *Politics & Society* 35: 447-475.

Weitzer R. 2012. *Legalizing Prostitution: From Illicit Vice to Lawful Business*. New York: New York University Press.

Werbner P. 2006. "Vernacular Cosmopolitanism". *Theory, Culture & Society* 23: 496-498.

Zelizer V. 2005. *The Purchase of Intimacy*. Princeton: Princeton University Press.

Zheng T. (ed.) 2010. *Sex-Trafficking, Human Rights, and Social Justice*. London: Routledge.

African Asylum Seekers in Israel

New Political, Racial and Urban Borders

Lisa ANTEBY-YEMINI

As migration flows and globalization increase, movement of goods and people across transnational space, borders and security regimes (surveillance, closure of borders, criminalization and detention of "illegal aliens") limit more and more migrant mobilities. In addition, once migrants – documented or undocumented – have crossed geographical borders, they are confronted with internal boundaries constructed by the state and the host society, that exclude them; yet they also challenge, contest and sometimes succeed in crossing, bypassing or blurring them. These processes of marginalization and incorporation, inclusion and exclusion, often characterize non-citizen migrants in the West, where they are incorporated in some areas (labour market) and denied access to others (welfare, citizenship) through legal means and informal practices (Castles, 2000). Scholars of migration have often studied geographical border-crossings and smuggling, state constructions of "illegality" and their implications for migrant daily life (De Genova, 2002), or humanitarian management of undesirable migrants (Agier, 2011); however the increase in policing of external borders in conjunction with the production of internal racialized boundaries is still an understudied area (Fassin, 2011). We will endeavour to explore this question through the example of African asylum-seekers in Israel.

Since 2005, an increasing number of African asylum-seekers, mainly from Eritrea and Sudan, have been clandestinely crossing the Egyptian-Israeli border. With the help of Bedouin smugglers, they reach the until-recently porous Israeli border. Since the 1990s, non-Jewish migrant workers have also been entering Israel, almost exclusively through Tel-Aviv's international airport; some hold work permits (such as those from China, Rumania, Thailand or Philippines) in the sectors of care, construction or agriculture; others usually arrive as tourists, from Africa and Latin America, and find informal employment in domesticity and services. Today 48,000 are legally working in Israel and an estimated

106,000 are undocumented (either because their work visas or their tourist visas expired).[1]

Asylum-seekers who entered through Egypt in the past six years number close to 60,000 and their presence is heatedly debated in Israeli society. Some are detained but the majority obtain temporary protection and settle in different cities, usually concentrating in specific neighbourhoods and working as cheap – often exploited – labour. This results in new forms of co-presence in Israeli society which transform the social texture of several neighbourhoods, often giving way to xenophobic discourses and, more recently, to violent acts of racism against asylum-seekers. At the same time, one also witnesses humanitarian assistance within civil society, defending refugee rights and offering services through a network of local NGO's. The production of these new political, social and urban boundaries, demarcating this non-Jewish migration to Israel, show how processes of "illegality", racialization and exclusion go hand in hand with mechanisms of social inclusion and urban territorialisation for these African populations in the country. While several researchers are currently studying asylum-seekers in Israel, few articles have yet been published on their life in Israel (Anteby-Yemini, 2009b; Sabar, 2010; Yacobi, 2010; Willen, 2008). We suggest examining in this article how, despite tightening geographical, political, legal, social and spatial borders, African asylum-seekers challenge these boundaries and find ways to cross them as well as erect new ones within the host society.

We will begin looking at the geographical, political and legal borders erected to exclude African asylum-seekers from entering Israeli territory and society. The recent and rapid evolution of control and closure (surveillance and border patrolling, construction of a "smart fence", detention centers) to manage this new migration flow parallels logics already used towards economic migrants in Israel; however additional challenges are complicated here by the asylum-seekers' clandestine entry through the land border with Egypt as well as Israel's international obligations as signatory of the Geneva Convention on Refugees (1951).

A second part will focus on the religious, cultural and racial borders that differentiate asylum-seekers from the Israeli population. Dynamics of social incorporation (through the workplace, schooling, interaction with NGOs, claims-making) at the same time as processes of social exclusion as "Others" are at play. Simultaneously, asylum-seekers reconstruct identities as Sudanese and Eritrean diasporic communities

---

1 Source: the Israeli Population, Immigration and Border Crossing Authority (https://www.piba.gov.il/PublicationandTender/Aug.2012).

that produce new identity boundaries as minority rights-claiming groups in Israel.

In a last section, we will concentrate on urban borders and territories where asylum-seekers concentrate in specific marginalized areas, leading to violent conflicts with locals, yet are also incorporated in urban space by constructing places of sociality and becoming residents of Israeli cities.

The material presented here is the result of several fieldwork periods in Israel during 2008-2012 where interviews were conducted with African asylum-seekers from Congo-RDC, Ivory Coast, Eritrea and Sudan and participating observation in shelters, cybercafés, churches, at social gatherings and demonstrations was carried out.[2] However, the experiences presented here should not be seen as representative of each group, because of large divisions inside each migrant community (along gender, religious, ethnic, tribal, regional and class division lines and motives of migration and status in Israel).

Geographical, political and legal borders

Crossing the border into Israel signifies, crossing geographical borders but, for the asylum-seekers, also crossing political borders between the "South" and the "North" and political regimes, from "dictatorship" to a dreamed "democracy". However, the recent construction of a "security fence", aiming at deterring African migrants as well as terrorist groups entering Israel through the Sinai desert, has put a new physical obstacle on the until now porous Egyptian-Israeli border. In addition, new restrictive measures of policing and control have added additional politico-legal barriers to the presence of asylum-seekers on Israeli territory.

Border-crossings and border-closures

As the number of African asylum seekers increased since 2007, the government decided in 2010 to build a "smart fence" on the Israeli-Egyptian border to deter this new migration flow that took spectacular proportions: in 2007, 5,383 migrants crossed the southern border of Israel clandestinely; in 2008, 9,142; in 2009, 4,437; in 2010, 14,746; in 2011, 17,201 and until June 2012, 9,071; since, the numbers have abruptly decreased, due to political unrest in Egypt and Sinai, closure of the border with the new fence and harsher treatment of the migrants by

---

2 Fieldwork in 2008-2009 was partly funded by the French National Research Agency (ANR) through the "MOFIP" project directed by C. Parizot.

Bedouin smugglers and Israeli authorities (see below).[3] Most are young, educated men but an increasing number of women, sometimes with young children, have been arriving, as well as unaccompanied minors; 63% come from Eritrea, 26% from Sudan (including Christians from South-Sudan and Muslims from Darfur and other regions), and in smaller numbers from Ivory Coast, Nigeria, Ethiopia, Ghana, Congo (DCR). The first arrived through the international airport in Tel-Aviv as early as the 1990s as tourists, pilgrims or economic migrants; since, increasing flows began coming through Egypt.

The main motives for migrating to Israel stem first and foremost from a feeling of insecurity (arrest, detention, deportation) in the transit countries the asylum-seekers go through (Sudan, Egypt, Libya), in particular after the tragic sit-in of 2005 in Cairo,[4] but also because of the precarious living conditions in these countries (no work permits, few schooling options, no social or medical rights and daily racism).[5] Israel thus appears as a new destination fro those fleeing religious and ethnic persecutions (Sudan, South-Sudan, Eritrea), unlimited military conscription (Eritrea), civil strife and conflicts (Congo-DRC, Ivory Coast) and violation of human rights. Christian and Biblical motives also play a role in the choice of migrating to the "Promised Land" (Anteby-Yemini, 2009a; Sabar, 2010), as a South-Sudanese preacher pointed out to me, after arriving in Israel in 2008: "it's written in the Bible that we will come here through Egypt, like the Exodus of the ancient Hebrews".

Some asylum-seekers travelled years before reaching Israel and lived for various periods in transit, such as Maurice[6] who fled Congo (DRC) to Mali and then to Algeria, from where he was expelled back to Mali; he then went to Cairo where he heard for the first time, through other migrants, about Israel and decided to join them. For others, this destination is already planned from home and can be reached in a few weeks, such as Geta's case, an Eritrean imprisoned and tortured in his country for deserting the compulsory unlimited army service; he managed to flee and planned to join his uncle who had already migrated to Israel. If in the past, however, narratives circulated in Cairo about Israel's high salaries, easily-obtained protection for asylum-seekers and supportive

3 A total of 62,792 migrants entered but 57,193 remained in Israel as of June 2012. Source: Israeli Population, Immigration and Border Crossing Authority (https://www.piba.gov.il/PublicationandTender/Aug.2012).

4 In December 2005 Sudanese asylum-seekers in Cairo organized a two month sit-in in the central Mahmoud Square to protest their treatment by the UN High Commissioner for Refugees. The Egyptian police dispersed them, resulting officially in 27 victims but according to the migrants over 200 died (Harrell-Bond, 2006).

5 See for example Le Houérou (2005) and Grabska (2006) on racism in Egypt.

6 All names have been changed.

NGOs, today this route is becoming less attractive, because of escalating border violence and Israel's new repressive policies.

Itineraries have also been quickly evolving and more organized passages start now from the countries of origin, rather than from Egypt only, through a wide network of Bedouin smugglers working in Eritrea or Sudan (sometimes with the help of locals). Crossing the Sinai desert, an ancient trafficking route for goods dominated by an informal economy of Bedouin smuggling, has also become highly dangerous – with increasing trafficking in human beings and trafficking of organs; in fact, since 2010, mostly Eritrean asylum-seekers are being detained, tortured and sexually abused by Bedouins in Sinai who ask for large amounts of ransom money to release them,[7] blurring the line between human smuggling and trafficking and giving rise to new forms of criminality and border economy. In addition, when the asylum-seekers reach the border-zone with Israel, they are in danger of being arrested, detained, deported to their home country (which often spells detention and even death penalty) or shot by Egyptian soldiers (Anteby-Yemini, 2008). This excessive brutality at the border sadly translates into a rising death toll. Other dramas occur when families are separated, some arrested on the Egyptian side and others managing to cross into Israel, often wounded by the Egyptian soldiers' bullets or from the barbed-wire. Border-crossings thus leave scars on the body and the mind, embodied in often terrible ways; women who become pregnant from the rapes they endured in Sinai are but one example of how the experience of the border can be rooted in the very flesh – and here the womb.

Even though they are aware of these dangers that may cost their life, many asylum seekers claim they have no choice to migrate elsewhere and continue crossing into Israel, seeking protection, as Sarah, who arrived from Eritrea in 2011, told me: "We have not come to work; we came to seek refuge; all we want is protection". However, the near-completion of a 220 km fence along the Israeli-Egyptian border (of which 205 km is already built as of Sept. 2012) erects a new obstacle to this migration flow with its 5 m. high steel grid, cameras, radars and motion detectors. This reinforcement of the border reflects global trends towards physical closure of national borders to combat immigration and terrorism (Shamir, 2005) but raises new legal and ethical dilemmas for the Israeli State confronted with both containing non-Jewish African asylum-seekers and respecting the Geneva Convention on Refugees it signed. With asylum-seekers now begging at the fence, Israel must decide if it has a moral and international obligation to let them cross or

7 See *The Dead of the Wilderness: testimonies from Sinai Desert*, Hotline for Migrant Workers report, 2011, Tel-Aviv.

if it will not take responsibility for migrants outside its borders, as Israel's Attorney General stated.[8] However, if asylum-seekers' lives are in danger in Egypt (which does arrest, detain and often deport them – difficultly qualifying as a "safe country"), Israel would not be abiding to its international commitment to non-refoulement of asylum-seekers, which it has on the whole respected until now (despite isolated cases of "hot returns" i.e. immediate deportation to Egypt).

Blurring politico-legal borders: "refugees" vs. "infiltrators"

Asylum-seekers in Israel represented a new and fairly unknown category in non-Jewish migration flows to Israel until the last decade. A new management of asylum-seekers radically changes the State's implication, which until recently let the UNHCR handle claims, interview asylum-seekers and recommend their files to the Israeli State, which made the final decision. With the creation in 2009 of the *Population, Immigration and Border Crossings Authority* of the Israeli Ministry of Interior, the State is now in charge of processing all asylum claims, determining refugee status and issuing visas. How Israel resolves the question of the asylum-seekers' entry, stay, status and rights reflects a current lack of asylum policy but also the confusion it results in. Indeed, Israel treats asylum-seekers in an incoherent way, which ranges from policing measures characteristic of a regime of repression to humanitarian measures of a regime of protection, comparable to contemporary trends in Europe (Agier, 2011). Because of the legal, political and administrative chaos or what Willen (2009) calls the "governmental turbulence" which characterizes these forced migrants' situation in Israel, they often end up living in a "legal limbo" (Kritzman-Amir, 2009).

If some are detained (see below), the majority of African asylum-seekers (Sudanese and Eritreans) are released "on condition" and obtain collective temporary protection *prima facie* (i.e. by being identified as nationals of these two countries) with a renewable visa (varying from 1 to 3 months) but no work permit; however, they are denied access to refugee status determination (RSD) procedures because Israel is reluctant to recognize refugee status (according to the Geneva Convention) which would lead to permanent residency and claims for family reunifi-

---

8 In September 2012 a group of 21 Eritrean asylum-seekers begged Israeli soldiers, through the new steel fence, to let them enter Israel; the soldiers gave them water and only allowed two women and a child to cross into Israel for humanitarian reasons; see G. Cohen "Police Bar Israeli Doctor's NGO from Accessing Eritrean Refugees Trapped at the Border", *Haaretz*, 06.09.12.

cation.[9] Other nationals (Nigeria, Ghana, Mali…) who do not obtain temporary protection are threatened of deportation if their asylum claim is rejected and can be detained for long periods, if they refuse to return to their home countries or to identify themselves, lack travel documents or are not recognized by the country's officials as citizens (as is the case of several Ivoirians). Because of this "condition of deportability", i.e. the possibility of deportation (De Genova, 2010) asylum-seekers live in legal vulnerability, including Eritrean and Sudanese with "conditional release" visas that only "defer deportation". Furthermore, nationals from Ivory Coast face "deportability" since their protection ceased in 2010 and South-Sudanese have been brutally experiencing it since they are being deported as of June 2012 when their new country was deemed "safe" by Israeli authorities. This allows the State to control the "temporary protected" population through a status that can be abruptly ceased and by creating legal borders with the status of citizen, that are almost impossible to cross, since there is no access to regularization in Israel (except through marriage with restrictions) and no *jus solis*. Only a minority of Muslim asylum-seekers from Darfur (450) received in 2007 temporary resident status, as a humanitarian gesture, and enjoy the rights associated with it (access to welfare, work permit).

However, because of differential treatment according to country of origin and date of arrival, new categories categorize individuals inside these migrant communities, creating borders and hierarchies between statuses (asylum-seeker, "closed file", refugee, recipient of a temporary protection, temporary resident) but also between nationalities (Eritreans and Sudanese vs. other asylum-seekers) or ethnicities, such as the Fur (from Darfur), who were given temporary residency because of a religious and historical affinity toward "genocide survivors" (Anteby-Yemini, 2009a; Willen, 2008). Yet the hyper-mediatization of Darfur refugees created "hierarchy in suffering" among asylum-seekers and often led to the instrumentalization of this group by local NGOs (Willen, 2008).

Those waiting for RSD and those with temporary protections do not receive any financial assistance nor medical insurance and lack social rights. By excluding them from social and political rights, Israel wishes to discourage settlement of non-Jews, joining the same logic of exclusion regarding non-Jewish migrant workers (Kritzman-Amir, 2009). The consequence of this paradoxical policy of repression and humanitarian assistance is the emergence of rightless non-citizens, that cannot be expelled (according to UN's High Commission on Refugees' guidelines

---

[9] Israel's rates of recognition of asylum seekers as Geneva Convention refugees scores one of the lowest in the world (0.5% in the past years); only 190 refugees have been recognized since 1948, and only 1 was approved in 2011 (*Refugees' Rights Forum*).

for Eritreans, Sudanese and Congolese as of June 2012) but with no possibility to pursue their migration – leaving them between political borders of rights and territorial borders of the state. Similarly in the West, decline in asylum-granting is coupled with an increase in discretionary forms of humanitarianism (Fassin, 2011) which come to replace political asylum, often leading to precarious sub-statuses replacing the protective Geneva Convention refugee status (Agier, 2011; Legoux, 2003).

Furthermore, as Israeli civil society and asylum-seekers themselves put forth their identity as "refugees" and construct an image of vulnerable victims, official discourse qualifies them as "infiltrators", reflecting a criminalizing definition which designated in the 1950s Palestinians and Jordanians who had entered clandestinely Israeli territory. Sometimes officials even go so far as coining new expressions, like "work infiltrators" (*mistanenei avoda*), trying to amalgamate in public opinion asylum-seekers and undocumented economic migrants (Anteby, 2009a). Thus, the state is intentionally blurring the borders between asylum-seeker and criminal, "alien with a temporary protection" and "illegal enemy alien", joining other world-wide criminalization of asylum-seekers (Agier, 2011; Castles, 2000). However, the socio-political process of "illegalization" (De Genova, 2002) of African asylum-seekers subtly covers up the very paradox that the vast majority are legally present in Israel, since they hold renewable "conditional release" documents, issued by the Israeli Ministry of Interior, allowing them to remain in Israel! De Genova (2002, p. 439) defines migrant "illegality" as a spatialized socio-political condition, central to ways migrants are racialized as "illegal aliens" in nation-state space. In this sense, the Israeli state's production of asylum-seekers' illegality intentionally shifts the category of "legal" to "illegal", since most asylum-seekers are by no means illegal (in juridical terms) nor undocumented, as opposed to a minority from countries other than Sudan, Eritrea or Congo, or to foreign workers who may well be undocumented…

In addition, recent government measures point to the tightening of these politico-legal borders. In the past years, asylum-seekers who crossed the border were apprehended by Israeli soldiers who would transfer them to the Prison services in charge of detaining anyone who entered the country by "infiltration" (i.e. from an enemy country or who crossed a country hostile to Israel). In response to the massive flow, a makeshift detention centre in the desert, made up of tents and prefabricated cubicles, was created in July 2007 adjacent to the Ktziot prison, close to the Egyptian-Israeli border. With a capacity of 2000, this facility was quickly over-occupied and asylum-seekers were often left off by the soldiers and made their way to Israeli cities, or were released after short periods of detention, if they proved to be Sudanese or Eritrean nationals. However, as of June 2012, any "infiltrator" is systematically detained, according to the new Anti-Infiltration Law, up to three years without judicial review – or indefinitely if he or she is a citizen of an enemy state, such as Sudan. The construction of a new detention center near the Egyptian-Israeli border is also being carried out to implement this new confinement policy, echoeing the recent spread of detention centers for asylum seekers and undocumented migrants in Great Britain, France, Malta or Australia (Agier, 2011; Shamir, 2005). This, no doubt, will have implications for mobility inside Israel, where some asylum-seekers will move freely while others will remain detained, embodying different borders of statuses and spaces. The production of these identity categories, reflects a global tendency, with temporary statuses that classify migrants as tolerated, detained, documented, or undocumented and corresponds to various modalities of recognition and assistance that includes or excludes them (Agier, 2011). This heightening of control and policing of external physical borders and internal political-legal borders is also matched with the production of internal social and racialized boundaries (Fassin, 2011).

Racial, cultural and religious boundaries

The recent African migration represents the figure of a new "Other" in the Israeli migratory landscape, alongside the foreign worker. As elsewhere, asylum-seekers and economic migrants "by demultiplying the presence of 'Others', the plurality of 'Others' and the plurality of their forms of presence, reframe the question of otherness at the heart of middle-eastern societies, while simultaneously demultiplying the angles of this alterity" (Bensaad, 2009: 21). In Israel, African asylum-seekers are often portrayed as "Blacks", "Africans", "non-Jews" and ultimate "Others", demarcated by clear racial, cultural and religious boundaries. However, social boundaries are sometimes crossed, shifted or dissolved when asylum-seekers and their children participate in socially interactive areas, such as school workplace, civil society or community-based activities.

"Othering" African asylum-seekers

If Israeli authorities have been increasingly depicting the asylum-seekers as criminalized "infiltrators" that should be incarcerated or expelled (see *infra*), this image has also been reinforced by the rhetoric of right-wing politicians, who consider them a demographic threat to the Jewish and Zionist identity of the country. The Israeli Minister of Interior, Eli Ishaï, went so far as to declare that African "infiltrators" pose an "existential threat" to Israel.[10] The discourse of certain political and religious figures, speaking of African asylum-seekers who spread diseases, increase rates of criminality by stealing goods or raping women, and render public space insecure, has in May 2012 reached an unprecedented peak, with the outrageous phrase of a Knesset Member, Miri Regev, who defined African "infiltrators" as "a cancer in our body".[11] This use of a discourse on migrants as a "societal illness", comparable to cancer, is also found in anti-immigrant rhetoric in Europe (De Genova, 2010), but takes a particular twist in Israel, where the collective memory as a Jewish country that should welcome persecuted "Others" clashes with the recent hostility towards these racialized "Others" that threaten this very same Jewish identity.

If in the past, foreign workers in Israel have also been perceived as a threat to the national identity, cultural homogeneity and Jewishness of the state (Semyonov and Raijman, 2004), asylum-seekers seem today to be even more regarded as outsiders, in the language of government officials and in certain xenophobic expressions of public opinion.

One example is how asylum seekers are depicted as "religious Others", especially since it is the first time Israel experiences such a mass influx of African Muslim migrants (mostly from Darfur but also from Eritrea) in addition to Christians (from Eritrea and South-Sudan). If some politicians and rabbis have highlighted that these asylum-seekers are a security threat because they come from an enemy-state (Sudan) and can easily associate with Islamic radicals, the asylum-seekers themselves do not put forth their Muslim identity; in fact they entertain tense relations with Arab Israelis. Some Sudanese mentioned to me their fears of becoming friendly with Arab (Muslim) Israelis, who are hostile to them for having chosen to seek refuge in Israel and perceive them as "traitors". They rarely visit the local mosques,[12] except for major holidays, such a the end of Ramadan when some pray at Al-Aksa in Jerusalem. Nonetheless, a few hundred Sudanese work and live in Arab villages in the North of Israel and others rent apartments in Arab-populated areas of Jaffa or East-Jerusalem.

As "racial Others", African asylum-seekers' ethnic visibility and racialization processes are complicated in the Israeli setting by the presence of the Ethiopian-Israeli population which challenges the boundaries

---

[10] D. Weiler-Polak, "Eli Yishai: Infiltrators Pose Existential Threat to Israel", *Haaretz*, 22.11.2010.

[11] See E. Tsurkov (2012) in the Hotline for Migrant Workers report on racial incitement against African asylum-seekers in Israel.

[12] Similarly, one notes little contact between Christian migrant workers and local Arab Christians in Israel (Kemp and Raijman, 2003; Sabar, 2008).

of solely racial difference lines. Close relations between the two groups are still rare, since a number of Israelis of Ethiopian origin refuse to associate themselves with the asylum-seekers. Roni, 20, born in Israel, claimed: “I don’t want to become close with them; we have nothing in common”. However, Marsha, 31, who immigrated to Israel when she was small, is more sympathetic: “When I hear the Eritreans’ stories, I remember how my family fled to Sudan and how hard escape journeys can be”. Another young Jewish man of Ethiopian origin adds:” I don’t want to have anything to do with African refugees, especially the Sudanese who were so cruel to my people in the refugee camps in Sudan, on their way to Israel”.[13] Interesting working relations develop, however, between African asylum-seekers and Israelis of Ethiopian origin. Esther, who immigrated from Ethiopia in 1991, owns with her husband an Ethiopian music and video store in south Tel-Aviv. She recently hired a young teenager, who arrived alone from Eritrea and understands Amharic, to help out in the shop; after a few days, she felt so sorry for him, she “adopted” him as a son, fed him and let him sleep at her house. On the other hand, Mark, an Israeli-born young man of Ethiopian origin, has an Ethiopian restaurant in Jerusalem. He hired an Eritrean asylum-seeker to clean and do the dishes but he says “Eritreans are mean”, “they’ve always been at war with Ethiopians” and in Israel “this hatred continues”. He pays him less than minimum wage but says he is so intelligent that one day he will probably open his own Eritrean restaurant right next door! In fact, Eritreans who began buying spices, traditional clothes and ethnic products in Israeli-Ethiopian shops are gradually opening their own... One does not therefore see the emergence of a shared identity as “blacks” among African asylum-seekers and Ethiopian-Israelis, because of the very different ways in which the racialized construction of each group is produced and because of the major distinction citizenship, identity and belonging play in differentiating the two populations; however, when it comes to racist discourse and violent attacks on the asylum-seekers, several Israelis of Ethiopian origin did hint at a common threat as “non-whites” in Israeli society, where racism can turn against “all blacks”.

Crossing social boundaries

African asylum-seekers in Israel are sometimes able – temporarily or permanently – to cross social boundaries and create new ways of belonging through interaction and participation in Israeli society at their workplace, in educational institutions, social services and activism.

---

[13] He is referring to Operation Moses, which secretly airlifted to Israel Ethiopian Jews who had walked from Ethiopia to Sudan in the hope of reaching Israel (Interviews with Israelis of Ethiopian origin, Tel-Aviv, May 2010).

Children of asylum-seekers' are incorporated in Israeli public pre-school and school system[14] and are building ties with local youngsters, learning quickly to speak Hebrew and acculturating rapidly to Israeli culture; a small number of unaccompanied teenagers have also been placed in Israeli boarding schools. Even though an exceptional decree was applied in the past to regularize the situation of children of foreign workers,[15] there is no present policy on the status of asylum-seekers' children – which means many remain stateless. Yet, interestingly, many African families choose to give Israeli first names to their children born in Israel, some explaining they hope to stay in the country. This social strategy, in their eyes, may somehow give their children chances to be included in Israeli society, in the same way perhaps as hanging Israeli flags in asylum-seekers' shops and homes can be seen as a instrumental social practice of belonging.

The workplace is also a space of interaction with Israelis and learning Israeli cultural habits and the Hebrew language. Bernard, from Ivory Coast, says he is good friends with Israelis in the restaurant he works at: "We often joke and eat together before the restaurant opens". Richard, from South-Sudan, says the Israeli owner of the small supermarket where he works often invites him to family gatherings and meals at his house. Thomas, once a biology student in Eritrea, claims that since he arrived in Israel he has been reading books about Jewish history, the Holocaust and knows all the Jewish holidays from his daughter's school.

However, it is perhaps gaining social acceptance and legitimacy in Israel that represents the most difficult boundary to cross. As the first massive flow of asylum-seekers streamed into Israel through Egypt in the summer of 2007, spontaneous mobilization of citizens, students and youth movements, in a mood of public empathy and solidarity, found shelters, donated food and collected funds. Civil society played a central role in stressing the humanitarian duty of Israel in welcoming African refugees (Anteby-Yemini, 2009a; Willen, 2008; Yacobi, 2010) and support committees, such as the *Committee for the Advancement of Refugees from Darfur*, were created by Israeli activists and academics; more recently, a number of NGOs have been instrumental in offering humanitarian support, monitoring asylum-seekers treatment in Israel, pushing for legislation to defend their rights and filing petitions against violation of their basic human rights.[16] In addition, the continual struggle

---

14 Regardless of their status, all children are schooled in the Israeli public system according to the compulsory education law. Some municipalities have however refused to enroll asylum-seekers' children (see next section).

15 See Kemp, 2007.

16 The *Refugee Rights' Forum* includes most bodies working in this area: Assaf, Physicians for Human Rights-Israel, Hotline for Migrant Workers, Kav le'Oved, As-

of left-wing Members of Knesset and public figures, who support the causes of asylum-seekers, also legitimates their presence in Israel.

Claims-making by the asylum-seekers can also be seen as a strategy of breaking social boundaries and obtaining public recognition. As excluded non-Jewish "infiltrators", asylum-seekers, as well as Israeli activists and local NGOs, endeavour to construct an identity not of distant strangers but of close similarity with Israelis, by invoking suffering and persecutions in Africa while establishing a parallel with Jewish history and the "refugee" past of many Israeli citizens. In this context, asylum-seekers, particularly from Darfur, do not hesitate to compare the horror of the genocide in Sudan to that of the Holocaust in Europe (Anteby-Yemini, 2009; Sabar, 2010; Willen, 2008; Yacobi, 2010), wearing for example t-shirts with "Darfur-genocide-Shoah" inscribed in English, Hebrew and Arabic, speaking about their traumatic experiences, and creating associations to make their fate known, such as the *Darfur Sons Union in Israel.*

Simultaneously, asylum-seekers also call upon a more global discourse on asylum. Migrant workers in Israel have since the 1990 set up associations to defend their rights and promote the cause of undocumented migrants in the public and political spheres, through invoking a global language of human rights, conveyed by local NGOs and their lobbying in the Knesset (Kemp and Raijman, 2004; Sabar, 2010). Today, asylum-seekers are creating their own organizations, such as the *African Refugee Development Center*, to defend, promote and claim their rights as "refugees" and integrate the international human-rights discourse on asylum; they demand access to RSD procedure, access to social and medical rights, in a way contesting the boundaries of citizenship and humanitarianism. One therefore observes, among asylum-seekers and NGOs in Israel, on the one hand local claims, linked to a rhetoric as victims of persecutions, specifically aimed at an Israeli audience, and on the other, global claims that join a transnational discourse on refugee rights (Anteby-Yemini, 2009a).

Community-building is also a way of overcoming social borders by erecting new symbolic borders of difference and producing ethnic and cultural identities. As asylum-seekers begin to settle, new technologies of communication (cell phones, internet) play a considerable role in the construction of local and transnational diasporic identities. Satellite TV also has a central function as source of information and link to an imagined community, such as the French-speaking channel TV5 for the

---

sociation for Civil Rights in Israel, Amnesty International, African Refugee Development Center and the Program on Refugee Law of the Tel-Aviv University Law Faculty.

Congolese, the channels of the Arab world for the Sudanese or Eritrean TV. Furthermore, various cultural associations, mutual help groups (rotating credit or funeral groups), religious structures and even football teams have been established. One also witnesses an increasing politicization among asylum-seekers organizing not only in ethnic communities and national communities (e.g. the *Association of Eritrean asylum-seekers*), but also in pan-African associations (ARDC), as well as political diasporic activism, exemplified by the opening in 2008 of a branch of the *Sudan Liberation Movement* (SLM) in Tel-Aviv.[17]

Urban borders and territories of belonging

Let us now examine how, in spite of being marginalized in poor areas, living in sub-standard conditions (sometimes even homeless) and experiencing tense relations with locals, African asylum-seekers also manage to appropriate public spaces and create territories of belonging that both incorporate them in the city while creating new urban frontiers.

Marginality, racism and neighbourhood frontiers

Because there exist no hosting centres nor any right to housing for asylum-seekers in Israel (as opposed to special hostels in France or Germany, for example), lodging is a constant problem and many remain homeless until they find a place to live. The Levinsky Park in Tel-Aviv has become the main area where homeless asylum-seekers sleep and where a soup kitchen serves free meals: Roy, a student from Eritrea, has just been released from the detention center near the southern border and has slept there for the past 3 days; Benjamin, from Sudan, who lost his job and has no more money to share the rent, has slept there for the last week. Ali, who is in Israel since 2009, from Darfur, comments as we walk through the park: "Israelis pay taxes to have a nice park and not a place where Africans sleep; the government of Israel needs to do something about this. It's shameful for us too".

The veteran asylum-seekers' community in Israel, however, constitutes a privileged space of integration for the newly arrived. Often crossing the border with their Egyptian cell phones, newcomers rapidly contact friends or relatives who offer hospitality and provide emergency financial support during the first weeks in Israel. In 2007 NGOs operated emergency shelters that are now only available for the most vulnera-

---

[17] The opening of this branch, that has been rather dormant, provoked an uproar in Sudan when the leader of the rebels in Darfur, Abdel Wahid Al-Nur, congratulated this initiative in Tel-Aviv; he was accused of being an agent for Israel and of not supporting the Palestinian cause anymore (see "Rebel commanders and IDPs support Israel office", *Sudan Tribune*, 03.03.08).

ble groups. After a varying period, newcomers usually share apartments with other asylum-seekers. Housing is often decrepit, landlords do not bother fixing dysfunctions and living space is overcrowded, with sometimes 6 or 7 people sharing two rooms. Those who earn enough move out, like Lema, an Eritrean studying at an Israeli university who lives alone, or Jim, from Senegal, who works several shifts in restaurants and shares a flat with a man from Ivory Coast he met when they were both detained and with an Eritrean who works in the same restaurant as he does. Sometimes couples form, such as Paul, an Ivoirian who put up a woman from Ivory Coast for several weeks and ended marrying her after a few months.

At first, many asylum-seekers began settling in poor and neglected neighbourhoods of South Tel-Aviv, where different waves of migrations have followed one another: Oriental Jews in the 1950s, Russian Jewish immigrants in the 1970s and 1980s and Asian, South American and African migrant workers since the 1990s. Spatial logics of residence made the southern quarters of Tel-Aviv become attractive for asylum-seekers because of low rents, employment opportunities, proximity to NGOs and services to foreign communities. However, in February 2008, restrictions in residence and work permits compelled the majority of asylum-seekers to settle north of Hadera and south of Gadera, excluding the Tel-Aviv region to prevent a "humanitarian crisis", as the government put it. This policy forced many to move to peripheral towns. However, some municipalities try to exclude them from urban space by restricting access to housing or schools. For instance, in November 2008, the mayor of Eilat declared he could not allow an additional 2000 Sudanese asylum-seekers to settle in the city because the social, medical and educational services were overwhelmed, with no extra funds from the State to accommodate this new population;[18] the asylum-seekers' children already in the city were being schooled outside Eilat, and after NGOs filed a petition to end this segregation, they were finally integrated in September 2012 into the Eilat public schools. More recently, the municipality of Bnei Brak (near Tel-Aviv) launched a campaign against renting apartments to asylum-seekers, following a letter by rabbis condemning landlords resorting to this form of income. If the "refugee problem" had indeed become a municipal concern, just as it had regarding the migrant workers (Yacobi, 2010), the stakes of their residential patterns demonstrate how asylum-seekers become a negotiation tool between municipalities who each choose to manage the "problem" differently.

---

18 R. Sinai "Israel to expel 2000 African Refugees who fled to Eilat", *Haaretz*, 26.11.2008.

The outcome of the Hadera-Gadera policy was that many asylum-seekers decided to settle in Hadera, Ashdod, Eilat or Arad for example but their massive presence in small cities also initiated opposition in the local population.[19] In July 2009 this policy of geographical control on asylum-seekers was abrogated by the Supreme Court and residential patterns converged again towards the Tel-Aviv region. This simultaneously refueled South-Tel-Aviv residents' protests against renting apartments to "Africans", blaming them for increasing criminality and insecurity, epitomized by the accidental death of an elder person beaten by a drunken Sudanese in February 2010.

Since, hostility and resentment has grown among residents who complain of insecurity, crime and violence on the part of the "Africans" who "steal, drink and harass women". Reports of an Israeli woman raped by Eritreans and other alleged attempts add to this climate of intolerance towards asylum-seekers' presence. Shula, who lives in Neveh Shalom neighbourhood in Tel-Aviv, claims she is afraid to walk down her street at night, where dozens of Africans gather on the benches, drinking beer. Lenny, an elder man from Shapira quarter says he feels unsafe sitting in the park where "Africans sleep, eat, urinate and defecate". Signs such as "Go back to Sudan" hang on some buildings. Haim, a resident of Hatikva neighbourhood, concludes: "they've conquered our neighbourhood!".[20] These rhetorics clearly call on images of invaded territorial borders and endeavour to reclaim "national" boundaries through regaining control of "local" urban territory.

During the first months of 2012 several criminal acts by Israelis against African asylum-seekers culminated in a violent anti-immigrant riot on May 23rd 2012; Molotov cocktails have been thrown against asylum-seekers' kindergardens and businesses, Eritrean stores have been vandalized, apartments have been set on fire in Jerusalem and Ashdod and asylum-seekers have been physically attacked.[21] Mulu, an Eritrean

---

19 Demonstrations took place in Arad against "African refugees" (estimated in 2008 between 700 to 1,000) and the competition they represent on the labour market, by accepting quasi-slavery conditions and low wages in the hotel industry of the Dead Sea (B. Moscona-Lerner and M. Grinberg "Arad: refugee city", *Maariv*, 17.07.2008). Similarly, in Shapira neighbourhood of Tel-Aviv, the residents' committee protests regularly against the concentration of "African refugees" who burden the social and educational services of their neighbourhood (Y. Haï "In the Shapira neighbourhood one does not want Africans", *Haaretz*, 07.08.2008).

20 Interviews, South Tel-Aviv, December 2011.

21 For a detailed account of this series of "hate crimes" see E. Tsurkov (2012), report for the Hotline for Migrant Workers. In June 2012, close to 100 Sudanese were removed by Israeli police from Kfar Manda, an Israeli Arab village in Galilee, after locals against the presence of the asylum-seekers aggressed them and injured 15

who has been in Israel since 2009 and owns a cybercafé in Jerusalem with his uncle, reflects: "Two apartments were burned in the past year; I'm afraid next time it will be mine. Sometimes, I think it was a mistake to have come to Israel and I would have been better to stay in Eritrea... Israelis insult me, tell me to go home to my country. I wish I could go somewhere else but I have no choice but to stay".[22]

These recent outbursts of racism, nationalism, xenophobia, fear and sense of loss of control are also voiced against migrant groups in Europe (Cohen, 2006). The new Right often uses this same discourse of alleged threats from ethnic minorities swamping the host society; in addition, disadvantaged groups usually share urban space with migrants: these "ghettoes of the disadvantaged" become sites of racist violence and extreme-right mobilization when migrants and locals compete for resources (Castles, 2000, p. 77). Israelis have already perceived foreign workers as socio-economic competition (Semyonov and Raijman, 2004) but the growing opposition to asylum-seekers now reaches new violent forms in contested spaces of cohabitation where the sudden surge in numbers lead to spreading African "urban territories".

Urban incorporation and social territorialisation

Despite producing urban marginality, asylum-seekers also show specific practices of the city that give them a new form of presence in urban space, through appropriating public places and creating territories of belonging. In the neighbourhoods they concentrate in, asylum-seeker invest certain places, such as parks and playgrounds, lawns and street benches, that rapidly become spaces of urban sociability; they are also specific streets where they wait for day-jobs. Those without documents (if their "conditional release" visas expired, their temporary protection ceased or their asylum-claim was rejected) live in fear of being arrested, detained and possibly deported, and experience urban space quite differently. They try to be the least visible in public space, particularly in face of the constant surveillance and identity controls by special units of the *Population, Immigration and Border Crossings Authority* in charge of arresting and deporting undocumented aliens. However, the majority of asylum-seekers have documents and cannot be deported (even though this can change), which explains why, for instance, hundreds of Sudanese and Eritreans are sleeping or chatting on the lawns in front of the new police station in Tel-Aviv's Levinsky Park... Their visibility is also noticeable in other parts of the city: they bike or take buses to work,

cf. H. Shaalan "15 Hurt in Brawl between Africans, Kfar Manda Residents", *Ynet News*, 14.06.2012.

22 Interview, Jerusalem, July 2012.

shop at local markets, and stroll along the beach but are most often seen cleaning streets and offices or working in restaurant kitchens and hotels. Depending on their visa type, some can work legally, others not (even though the Ministry of Industry, Trade and Labour is not strongly implementing sanctions on employers of asylum-seekers without work permits). These jobs are mainly found in Tel-Aviv, as in other global cities which have a high demand for low wage services (Sassen, 2011), leading in many cases to economic exploitation, labour restructuring and new inequalities.

A growing number of Eritreans and Sudanese have opened "ethnic businesses", especially in Tel-Aviv but also in Jerusalem and Eilat, such as restaurants, cybercafés, bars, nightclubs, hairdressers, groceries, photography shops, celebration halls, phone and computer stores, and informal child-care structures (Anteby-Yemini, 2009b). These shops are often decorated with Israeli flags, inscriptions in Arabic or Tigrinya, pictures of Barak Obama, posters of their home countries and resonate with Eritrean or Sudanese music... An informal flea market in Tel-Aviv's Neveh Shaanan pedestrian street is now dominated by African asylum-seekers buying and selling clothes, home appliances, bikes and portable phones. These ethnic businesses also serve a multi-ethnic clientele and create cross-national relationships (Eritreans eat in Sudanese restaurants, Chinese use Eritrean cybercafés, Ghanean women go to Sudanese beauty parlors). Churches and worship groups have also emerged (Sabar, 2010), creating new religious territories. Yet these places where men play dominoes, sip coffee, smoke *shisha*, and watch homeland DVDs on giant screens or where mixed audiences pray, represent first and foremost spaces of sociability which recreate symbolic Sudanese and Eritrean "ethnoscapes" in Israel cities. This reterritorialisation of places of migrant cultural practices produces new ways of belonging in Israeli space.

Thus, asylum-seekers are changing the human landscape of several neighbourhoods with an increasing African predominance, new languages (Arabic, Tigrinya and Sudanese dialects) and new religions (African Islam) – especially in towns that never received such populations (Arad, Sderot, Tiberias, Arab villages in the Galilee), operating a process of "globalization from the margins", to use Bensaad's (2005) expression referring to African migrations in the Maghrebi context. The asylum-seekers also consume goods and services, pay rents and can be considered as new consumers as well as cheap yet productive laborers, as Sudanese refugees in Cairo have been studied (Grabska, 2006); this demonstrates how marginalized urban asylum-seekers can nonetheless contribute to globalization processes in the city and be active in local economy.

Furthermore, just as migrant workers in Israel are concerned, practices of local incorporation by some municipalities go hand in hand with a state policy of exclusion of non-Jewish migrants (Kemp and Raijman, 2004). Certain cities, Tel-Aviv in particular, offers asylum-seekers various municipal services, such as the *Mesila* center that organizes extra-curricular activities for children of "refugees" and Hebrew and English courses for adults. A number of NGOs also cater the asylum-seekers' communities in the Tel-Aviv area (*Hotline for Migrant Workers*, *Kav le Oved*, clinic of the *Physicians for Human Rights-Israel*). The "humanitarian management" of asylum-seekers is thus carried out by the third sector and some municipalities that choose to provide services and assistance, playing a major role in integrating asylum-seekers in the city. One therefore witnesses the construction of an urban identity, similar to that of economic migrants, who became "new Tel-Avivians" incorporated in the city as "urban citizens" (Kemp and Raijman, 2004). Today, asylum-seekers, supported by local NGOs, demonstrate in public space, distribute flyers and participate in silent marches, performing urban visibility that differs from that of migrant workers – especially if undocumented – who prefer to remain invisible and not protest in public (Kemp and Raijman, 2003; Willen, 2007). Sudanese and Eritrean asylum-seekers ask for "urban rights" and "social residency" as *de facto* residents in Israeli cities, even though they are denied political and social rights at the national level.

Conclusion

We endeavoured to show that asylum seekers in Israel, in spite of the state's tightening of geographical, social, political and spatial borders, develop nonetheless ways of negotiating these boundaries and their denial of formal inclusion and rights; they also demonstrate patterns of spatial territorialisation by appropriating public space, by investing certain neighbourhoods and urban territories, and by becoming urban consumers. In fact, today, the new African communities of asylum-seekers create their own religious, economic, political and cultural structures, that model them into refugee diasporas linked to other Sudanese or Eritrean communities in exiled and turning cities such as Tel-Aviv into additional places of refuge on the map of global African forced migration. It is especially interesting to observe how quickly these territorialisation and community-building processes take place for populations in transit, just as they occur in other countries, such as Egypt (Grabska, 2006). At the same time, distinct and rigid boundaries are being erected between Israeli citizens and "infiltrators", mainly by state production of "illegality"; processes of social exclusion and racist rejection of "Africans" are also growing, through local residents' discourse, protests and violent acts.

As for the Israeli government, it is leading a policy similar to the management of other non-Jewish migration flows i.e. discouraging settlement in Israel and using detention or expulsion as deterrents; but here is added a humanitarian logic regarding asylum-seekers and in some cases survivors of genocide, for whom control and repression go hand in hand with forms of assistance and protection, or what Legoux (2003:19) calls in Europe a policy of "charitable asylum" in reaction to the *de facto* presence of asylum-seekers who cannot be expelled. Yet, Israeli authorities avoid directly taking in charge the asylum-seekers and informally institutionalize their "humanitarian management" by handing it over to municipalities and civil society. If this results in an exclusion policy on the state-level and certain inclusion practices on the local level, this dissonance also generates differential marginalization, where some groups are more included (e.g. asylum-seekers from Darfur) or more excluded (asylum-seekers without protection) than others. Given the State's new measures against asylum-seekers (systematic detention, sanctions against employers and "deportability", as experienced by the South-Sudanese), the question that remains open is how these rightless non-citizens will be able to survive and succeed in crossing the social boundaries of Israeli society…

References

Agier M. 2011 *Managing the Undesirables: refugee camps and humanitarian government*. Cambridge: Polity Press.

Anteby-Yemini L. 2008. "Migrations africaines et nouveaux enjeux de la frontière israélo-égyptienne". *Cultures & Conflits*, 72: 75-97 (http://www.conflits.org/index13683.html).

Anteby-Yemini L. 2009a. "Les 'réfugiés soudanais' en Israël: discours, représentations, mobilisations". *Maghreb-Machrek*, 199: 71-83.

Anteby-Yemini L. 2009b, "De la traversée clandestine à la visibilité urbaine: demandeurs d'asile et réfugiés africains dans les villes israéliennes". *Méditerranée*, 113: 13-24.

Bensaad A. 2005. "Les migrations transsahariennes, une mondialisation par la marge". *Maghreb-Machrek*, 185: 13-36.

Bensaad, A. 2009. "Le Moyen-orient: un Carrefour migratoire entre conflits territoriaux et mondialisation des circulations". *Maghreb-Machrek*, 199: 9-22.

Castles S. 2000. *Ethnicity and Globalization*. London: Sage.

Cohen R. 2006. *Migration and Its Enemies: global capital, migrant labor and the nation-state*, Farnham: Ashgate Publishing.

De Genova N. 2002. "Migrant 'Illegality' and Deportability in Everyday Life". *Annual Review of Anthropology*, 31: 419-447.

De Genova, N. 2010. "Migration and Race in Europe: the transatlantic metastases of a post-colonial cancer". *European Journal of Sociological Theory*, 13 (3): 405-419.

Fassin D. 2011. "Policing Borders, Producing Boundaries: the governmentality of immigration in dark times". *Annual Review of Anthropology*, 40: 213-226.

Grabska K. 2006, "Marginalization in urban Spaces of the Global South: Urban Refugees in Cairo". *Journal of Refugee Studies*, 19 (3): 287-307.

Harrell-Bon B. (interviewed by F. Le Houérou) 2006. "Le drame de la place Mustapha Mahmoud au Caire", http://terra.rezo.net/article553.html, consulted 17/01/07.

Kemp A. 2007, "Managing Migration, Reprioritizing National Citizenship: Undocumented Labor Migrants' Children and Policy Reforms in Israel". *Theoretical Inquiries in Law*, 8 (2): 663-691.

Kemp A. and Raijman R. 2003. "Christian Zionists in the Holy Land: Evangelical Churches, Labor Migrants, and the Jewish State". *Identities*, 10: 295-318.

Kemp A. and Raijman R. 2004. "'Tel Aviv Is Not Foreign to You': Urban Incorporation Policy of Labor Migrants in Israel". *International Migration Review*, 38 (1): 26-51.

Kritzman-Amir T. 2009. "'Otherness' as the Underlying Principle in Israel's Asylum Regime". *Israel Law Review*, 42(3): 603-627.

Legoux L. 2003. "Nouvelle donne mondiale, nouvel asile", in M. Guillon and al. (eds.). *L'asile politique entre deux chaises: droits de l'homme et gestion des flux migratoires*. Paris: L'Harmattan: 13-24.

Sabar G. 2010. "Israel and the 'Holy Land': the religio-political discourse of rights among African migrant labourers and African asylum seekers, 1990-2008". *African Diaspora*, 3(1): 43-76.

Sassen S. 2011. "Global Migrations and Economic Need", in R. Smith (ed.), *Citizenship, Borders and Human Needs*. Philadelphia: University of Pennsylvania Press: 56-91.

Semyonov M. and Raijman R. 2004. "Perceived Threat and Exclusionary Attitudes towards Foreign Workers in Israel". *Ethnic and Racial Studies*, 27(5): 780-799.

Shamir, R. 2005. "Without Borders? Notes on Globalization as a Mobility Regime". *Sociological Theory*, 2 (23): 197-217.

Tsurkov E. 2012. *"Cancer in Our Body": on racial incitement, discrimination and hate crimes against African asylum seekers in Israel*. Hotline for Migrant Workers, report January-June 2012, Tel-Aviv.

Willen S. 2007. "'Flesh of Our Flesh': Undocumented Migrant Workers' Search for Meaning in the Wake of a Suicide Bombing", in S. Willen (ed.), *Transnational Migration to Israel in Global Comparative Context*, Lanham: Lexington Books: 159-184.

Willen S. 2008. "L'hyperpolitique du 'Plus jamais ça!': demandeurs d'asile soudanais, turbulence gouvernementale et politiques de contrôle des réfugiés en Israël ". *Cultures & Conflits*, 71 (3)

Yacobi H. 2010. "Let Me Go to the City: African asylum-seekers, racialization and the politics of space in Israel". *Journal of Refugee Studies*: 47-68.

Rendering People Illegal and Invisible

Illegal Migrants and Daily Life (United States, Spain)

Virginie BABY-COLLIN

The 'migration planet' (Simon, 2008), in a context of world mobilities, is not in fact an era in which everyone can become mobile (Urry, 2005). Even though the concept of free circulation for all is well-established for certain regions, the introduction of 'open frontiers' is nevertheless accompanied by greater rigidity at various levels: not only at regional borders (Europe) and at national frontiers (United States), but also inside countries. This is known as a 'de-nationalization' of migration controls, "*in, down, and out*", i.e. inland from the national frontier, downward to lower-level government agencies and outwards beyond exterior boundaries (externalization of controls) (Guiraudon and Joppke, 2001). The financial, technical and human resources required to carry out border controls have increased massively, as have the numbers of expulsions of illegal migrants.[1] Policies contribute to the creation of barriers between those who have the right to enter and those who do not, or who have only partial access: limited either in terms of time (visas) or in terms of rights (entry permits but not work permits, residence but not vote, etc.). This leads to a complex system of different statutes which can change over time, from one place to another, and even within families themselves. These restrictions on the people's ability to move create legal borders even in the towns where migrants live, to the point where the stigma of their illegal status seem to be etched into their bodies, whether it is partial and temporary, total and permanent. While, in certain areas, some migrants have freedom of movement and have legal status, others are excluded as 'undesirables' (Agier, 2008). This term is used to refer to politically constructed categories within a variety of specific contexts, evidence of a 'biopower' (a term coined by Michel Foucault) which, by the general use of technological control, can put

1 There is a large number of terms for designating illegal migrants: 'undocumented' (even though some have false papers), 'irregular', 'clandestine', 'illegal'. We prefer the first two terms which are more neutral for people who have not respected the rules, rather than the latter two used for official statistics on illegality and of clandestinity, terms that we question. 'Illegal' migrants are often more 'legal' in their daily life for the very reason that they have to avoid checks and the possibility of arrest.

individuals into categories for different forms of treatment. The creation of illegality, as a 'spatial socio-political condition' (De Genova, 2004, 161), provides arguments for protection of a controlled national territory that confirms national identity by expelling 'undesirables' and, at the same time, establishes its importance in terms of strong social and economic usefulness.

Although a certain amount of research has analyzed changes in migration policies and issues relating to regularization (particularly in terms of economic questions),[2] few studies question the way in which illegal migration has an impact on the daily life of those who are victims, or may become victims, or have been victims at some stage during their migration experience (De Genova, 2010; Coutin, 2007; Agier, 2012). To what extent do the creation of illegality and the more stringent powers for its repression affect migrants' spatial mobility and daily movements? They force migrants to develop *tactics* of manoeuvre and withdrawal that demonstrate the ability of the *dominated* to find their place in society, as opposed to the *strategies* of *dominants*, according to the classic categories (Certeau *et al.*, 1990). They lead to the development of invisibilization, through restricted and specific visibility, which reveals curtailed '*citadinities*' (i.e. ways of inhabiting urban areas based on practices and representations), that become stakes in biased negotiations in a social environment which is only partially accessible. Here I propose to analyse these citadinities using a long-term research project on the geography of Latin American migrants in several cities in the United States and Spain that have a large number of irregular migrants. The situation in both the USA and Europe is becoming much more difficult with tougher migrant controls inside their borders and at the border themselves, but I will also challenge the way in which certain factors, such as the relative number of irregular migrants and the period of clandestinity (higher and longer in the USA than in Spain), affect the way that migrants adapt to their irregular status.

Following a discussion of recent changes in the current policy that creates migrant illegality in both countries and the principal measures that have an impact on individual mobility, I intend to review the various adaptation and resistance tactics that irregular migrants develop when living in cities with immigrant populations and to focus in particular on their day-to-day mobility.

---

2 See Sumption, 2011; Hanson, 2009; Levinson, 2005; Myers, 2007.

Parallels and divergences in the creation of illegality in Spain and the United States

Irregularity: despite its banality, it can fluctuate in time and space

In 2010, 12% of the population in the United States and 14% in Spain were born in another country. While the proportion of foreign-born residents is similar in the two countries, the reasons behind their presence are very different: the USA is a country built on immigration, whereas Spain, for many years a country of emigration, has now become a country of immigration following its return to democracy and its entry into the European Union (1986).

In the United States, there are 11.2 million irregular migrants, representing 3.7% of its total population, 5.2% of its labour force and 28% of its foreign-born population.[3] Four-fifths of these are from Latin-America and 58% are Mexican). The last campaign for massive regularization, the IRCA reform, took place in 1986[4] and, although the *Comprehensive Immigration Reform Bill* was debated in Congress, the events of 11 September 2001 have resulted in the reinforcement of security policy, rather than its enactment and implementation.

In the European Union of 27 Member States (EU-27), it is estimated that the number of irregular migrants ranges between 1.9 million and 3.8 million in 2008, i.e. 0.4%-0.8% of the total population and 7%-13% of foreigners (HWWI, 2009). Nearly 20 massive regularization campaigns have taken place since the 1990s with 5 million persons receiving papers (Brick, 2011; Kerwin *et al.*, 2012). In the period 1996-2007, 50% of regularizations took place in Italy (1.2 million) and Spain (just over 1 million, and 548,000 in the 6th and last campaign in 2005 alone) (Levinson, 2005). In Spain, it is estimated that there are currently 400,000 to 800,000 undocumented migrants (Caritas, 2010; HWWI, 2009), representing some 10% of the foreign-born population,[5] and 1-1.5% of the total population. The European Project Clandestino Report calculates that 40% of migrants in Spain during the 2000s were illegal and some studies suggest this figure is as high as 50% for Andalusia and

---

[3] Compared with 3.5 million in 1990, 8.4 million in 2000 and 12 million in 2008, its highest level until the number stabilized at around 11 million.

[4] Under the *Immigration Reform and Control Act* (*IRCA*), 2.7 million people were received US citizenship and, of these, the large majority was Mexicans. Between 1986 and 2009, special regularization programmes (such as the *Cuban Adjustment Program*, which covered 405,000 Cubans) gave US citizenship to 3.7 million migrants (Kerwin *et al.*, 2012).

[5] According an ENI survey carried out in Spain in 2007, the figure was 8%.

Catalonia (CERES and CCOO, 2004; Gonzalez Enriquez, 2009, 8).[6] Latin Americans, whose numbers in Spain have increased recently and massively (they represent now more than 30% of the foreign-born population and far outnumber any other non-European group) are the group with the highest level of irregularity: 50% of migrants regularized during the last campaign in 2005. Many arrived during the 1990s, before the requirement for visas was progressively introduced for Latin American countries after 2000 so that it was still possible to arrive as tourists without visas. Once visas became obligatory for Bolivians in 2007, 70% of the Bolivian community in Spain (about 250,000 persons) were illegal (Gonzalez Enriquez, 2009, 27-28).

In Spain, it is possible to apply for regularization after three years of continuous residence registered with the *Padrón*,[7] on condition of the applicant has a work contract (which explains why the most recently arrived groups are also more likely to be illegal) whereas, in the United States, procedures are much less more uncertain and complex, migrants can remain undocumented for years, even when they have a job. To overcome this situation, migrants have to establish a strong legal case. Illegal status can vary over time and most labour migrants will, at some time or another, have been living without papers. Around 60% of irregular migrants on American territory arrived with a visa (usually as tourists) and became illegal by 'overstaying' after the expiry of their visa.[8] Even receiving an initial residency permit is no guarantee that a long-term permit will be available, because as it is based on having an employment, loss of that job will immediately make the migrant illegal again. The obligation to renew this permit during the early years before a full residency permit is issued, contributes to maintaining, if not actually increasing, the number of irregular migrants.[9]

---

6 Gonzalez Enriquez (2010) refers to the rule of irregular immigration.

7 The municipal *Padrón* is the city register for the Spanish population; registration is obligatory for residents and their access to basic rights (e.g. enrolment of children in schools). Registration in the *Padrón* provides proof of residence and must be produced by irregular foreigners when preparing their application for naturalization. It is updated each year with online access and is an exceptional resource for counting the population.

8 The other 40% crossed the US/Mexico border illegally.

9 There are restrictions on mobility for holders of US green cards and permanent residence permits. Residents can enter and leave the country, but if they stay more than one year outside the country, they lose their status; they must apply for an entry permit after an absence of more than 6 months, etc. Some green cards are subject to conditions (during the first two years). To obtain American citizenship, candidates must have lived in the USA for at least 5 years on a green card and spent more than 18 of the previous 30 months in the United States (see http://www.uscis.gov, the US Citizenship and Immigration Services site).

Thus, migrants live in uncertainly for many years, and this is all the more frequent in that American procedures are long and there have been no amnesties since 1986: many migrants have been illegal for nearly 20 years. By comparison, Spain, with the massive regularization campaigns in the 1990s and 2000s (6 in all) and much more flexible arrangements, has a much lower level of irregular migrants and shorter periods of illegal presence. Nevertheless, the recent introduction of tougher migration controls is having a greater impact, in both countries, on the daily lives of undocumented migrants.

Recent introduction of tougher migration policies

The events of 11 September 2001 created a watershed in essential security which had a heavy impact on migration policies, the repercussions of which have been felt throughout the world. As a Member State, Spain was committed to the EU's policy of greater security and closed frontiers supervised by Frontex, the European frontier agency created in 2004. Since 2007, the world economic crisis has given fuel to highly publicized media campaigns that seek to place responsibility – in part – for the crisis, and in particular in the job market, on immigrants. Focusing on clandestine immigrants as a source of danger, and thus a threat to national identity, national economy and national security, goes hand in hand with the introduction of tougher policies. In addition to tighter controls at national frontiers, which will not be discussed here, the higher level of expulsions suggests that, from now on, arrests now take place inside national territories as a result of detainment for minor offences or identity checks. In the USA, expulsion orders were issued for 165,000 irregular migrants in 2002 and 396,900 in 2011 (US-ICE, 2011). Of these, 97% were Latin Americans (who represent 80% of illegal migrants). In Spain, MIR (the Spanish Department of Security) ordered the expulsion of 10,616 persons in 2008, 13,278 in 2009, 11,454 in 2010.[10]

While Spain was strengthening control procedures in a relatively consistent manner throughout its territory, the United States published a

---

In Spain, after a period of legal residence (i.e. after the 3rd annual renewal of their residence permit and on condition that they are employed) – 2 years for some nationalities (including Latin Americans), 5 for others – migrants can apply for naturalization. For legal Latin American migrants, the procedure seems relatively fast.

10 In 2009, according to police union sources, 445,000 identity checks were made in Madrid; the same year, in Spain, of the 13,278 arrests of illegal migrants, 57% were 'qualified' (i.e. for infractions or crimes) and 43% 'non-qualified' (i.e. without infractions). In 2010, of the 54,000 non-qualified arrests of illegal foreigners, 6-7% led to expulsions (i.e. 29% of 11,454 expulsions). However, because the *Centros de Internamiento de Extranjeros* (CIEs) have limited space, the majority of expulsion orders are not executed (information supplied by the Ministry of the Interior (MIR)).

very high number of regulations at various levels of government (from local to federal including state legislatures) and the situation becomes extremely complex with policies often in contradiction from one state to the next.

In 2010, Spain passed new migration legislation which tightened the entry conditions and restricted the rights of immigrants[11] (as well as increasing sanctions, limiting rights to family reunion,[12] and extending the maximum period for retention in internment centres for foreigners called *Centros de Internamiento de Extranjeros* (CIEs))[13] and then, in 2012, removed access to free health services for irregular migrants (with the exception of pregnant women, children and emergencies). Circular 1/2010, issued by the Commissioner General for the Exterior and Frontiers and sent out to all Spanish commissariats, gave instructions for more identity checks in the street, with the possibility of 'preventive detention' and requiring them to be ready, around the clock, to execute immediate expulsion orders with systematic delegation of signature.[14] Arrests can be carried out on the basis of a simple identity check, and not just for felonies, and these persons placed in a CIE. Although the Ministry of the Interior denies the existence of *redadas* (raids), a term often used in the press and reports and has stated that the police is simply carrying out identity checks, as part of its responsibility for maintaining security and fighting against terrorism, the Unified Police Union (SUP) has denounced these security checks which are being used to track down undocumented migrants.[15] In 2009, leaks to the press refer to internal documents which fix weekly quotas for the arrest of irregular

---

11 Reform of the Organic Law 4/2000 on the rights and freedoms of foreigners in Spain and their integration into society by the Organic Law 2/2009 dated 11 December 2009.

12 'Family reunification' is limited to the nuclear family, children under 18 and a married partner or person 'with a loving relationship similar to marriage', and only parents over 65 are admitted. Applicants must have received the first renewal of their resident permit, have a work permit and produce proof of sufficient resources and lodgings to accommodate their family.

13 The maximum stay was extended from 40 to 60 days; after this, migrants must be released with a copy of their expulsion order.

14 Immediate expulsion orders can be issued if illegal migrants have 'no stable domicile but a temporary or transitory lodging', and they do not live with their family but with 'other persons' (Circular 1/2010, *Comisaria general de extranjeria y fronteras*).

15 A SUP press release (9 February 2010) denounced the obligation that police officers apply Circular 1/2010, on the grounds that "*massive and indiscriminate identification checks in public streets ... weaken citizens' rights and endanger the security of police officers*".

immigrants and specify which nationalities are to be targeted ("*Bolivians, not for the moment, limited capacity on planes*").[16]

In the United States, even though implementation of migration laws is a federal responsibility and delegated to the Department of Homeland Security (DHS)[17] by Immigration and Customs Enforcement (ICE),[18] the entry into force of Article 287(g) of the *Illegal Immigration Reform and Immigrant Responsibility Act* (IIRIA), once changes relating to the events of 11 September 2001 were incorporated, authorizes states, counties and local agencies (specifically local enforcement agencies (LEA)) to sign conventions directly with ICE.[19] Thus, local agencies can tighten controls on immigration and clandestine migrants on American territory; however, this has led to a number of irregularities, including arrests for minor infractions and incidents of racial profiling in order to raise the number of deportations, not the usual task of local police services established to maintain law and order (Rodriguez et al., 2010).

Meanwhile, since 2008, the *Secure Communities Program*, initiated by ICE, can be used to compare fingerprint records of persons placed in local jails with DHS files in order to facilitate the identification of irregular migrants (Waslin, 2011). In 2011, 41% of court jurisdictions had signed up for this programme (which is expected to cover the whole country by 2013) and this has led to the expulsion of 101,741 persons in 30 months. Concerned about inappropriate comparisons of files on individuals arrested for minor infractions, several states have publicly withdrawn from the programme, including Illinois in May 2011 and New York in June the same year.[20]

---

16 Cf. http://www.elpais.com/articulo/madrid/policia/fija/cupos/arrestos/papeles/barrios/elpepuespmad/20090216elpmad_3/Tes.

17 Established under the *Homeland Security Act* 2002, it merges 22 agencies and border control, customs, regularization and immigration services into a single administrative body. This was the largest single government reorganization of security and immigration services since World War II and its primary objective was to improve the efficiency of the policies for US security and against terrorism in response to threats from terrorists (Mittelstadt *et al.*, 2011).

18 ICE (www.ice.gov) was created in 2002 by merging the US Customs Service and the US Immigration and Naturalization Service (USCSINS) with their 20,000 employees.

19 In 2010, 71 jurisdictions have signed the Memorandum of Agreement (MOA): 13 states, 43 counties, 15 cities. There are three different formats.

20 Mr. Cuomo, Governor of the State of New York, announced his withdrawal from the programme on 1 June 2011 because "*there are concerns about the implementation of the program as well as its impacts on families, immigrant communities and law enforcement in New York*".

State legislatures passed 64 anti-immigration laws in 2010, and another 100 in 2011:[21] Arizona's SB1070 (passed in 2010) is thought to be the most repressive in the country: under its provisions, being an illegal migrant is against the law and is punishable by a prison sentence. The local police can investigate the status of "*anyone they stop, detain or arrest for a state or local offense if the officers have a* reasonable suspicion *the individual is an unauthorized immigrant*". Some cities[22] have passed *Illegal Immigration Relief Acts* (IIRA) which can cancel driving licences held by illegal workers, impose fines on landlords and employers of illegal migrants,[23] and require that English is the only language spoken in city stores. Following very lively debates on the question of migration, several local and state legislatures have, on the other hand, introduced policies promoting a variety of decrees and directives favourable to immigrants: issue of city identity cards,[24] guarantees of confidentiality for municipal employees relating to data on migration, encouraging use of other languages in government agencies.[25] The complexity of local structures and methods for implementing these laws, which are often in contradiction with other jurisdictions (state, county and municipal) demonstrates how extremely difficult it has become to navigate through the immense range of territories. Migrants need to have expert knowledge of local laws in order to find out where it is easier to live, drive, work and find lodgings and where a stroll in village streets can lead to expulsion from the United States.

---

21 According to the database developed by *Mother Jones* in 2012: http://www.motherjones.com/politics/2012/03/anti-immigration-law-database#database.

22 E.g. Hazelton (Pennsylvania) and Fremont (Nebraska), medium-sized towns with recent immigration.

23 Federal sanctions against employers of irregular migrants are foreseen since the *IRCA Act 1986* now that employers can use the online *E-verify* system to check employees' legal status by entering their social security number. This system has been undermined by clandestine migrants by using false social security numbers or numbers of others, even deceased persons (Meissner and Rosenblum 2009). The legislation is also ineffective because there are not enough inspectors and authorities are not interested in checking records.

24 New Haven (Connecticut), San Francisco and Oakland (California) provide city identity cards which maintain public law and order while allow people to avoid having to produce other documents to prove their legal status (e.g. driving licence) (Bloemraad and De Graauw 2012).

25 E.g. New York City's *Executive Order No. 41*, 2003, ensures confidentiality of individuals' data in city service departments; the city has set up multilingual services and a specialized policy unit, as part of its immigration service, which guarantees confidentiality on migrants' status.

Daily life of undocumented migrants: mobilities and spatial tactics

The introduction of tougher policies on controls and criminalization of irregular immigration has a profound impact on the daily life of migrants and their families. As they run the risk of expulsion, their rights as foreign citizens are reduced, to the point where it is almost impossible to have them respected, and their capacity to move around every day depends on how much latitude is allowed under local legislation and its implementation. Their survival tactics depend on their appreciation of their capacity to manœuvre and negotiate within the local application of state or federal regulations. To what extent are personal mobility and occupation of space affected by the legal or illegal status of migrants in these circumstances?

Restrictions on overseas travel

Inside the European Schengen zone, irregular migrants can cross borders freely but they cannot leave Fortress Europe (MIGREUROP & Clochard, 2009) because they may never be able to re-enter. In the United States, international travel, even for irregular migrants, was possible by crossing the American/Mexican frontier illegally until the 1990s but successive reinforcement of border controls makes this option less attractive and, consequently, has reduced the number of illegal journeys but increased the number of long-term migrants. Smugglers have raised their prices considerably now that border crossing are more difficult, while the presence of mafias and higher recruitment of border patrol officers make the whole process much more risky (Cornelius, 2005; Riosmena and Massey, 2012; Baby-Collin and Mercier, 2012).

Irregular migrants, because of their inability to travel, are thus forced to become long-term residents in the hope that they will one day obtain regularization. Legal papers are synonymous with the ability to travel and the possibility of at last returning to their country of origin. For irregular migrants, their inability to travel abroad is often linked to a reduced local mobility, due to new restrictions within their country of destination.

Geography of invisibility guided by fear

Tougher police controls within national territories lead to enforced invisibility. *Regimes of visibility* constructed by the dominant vary considerably, as do tactics to meet new situations at any given time (Bélouin *et al.*, 2009; Revue du MAUSS, 2004; Voirol, 2005; Lussault and Lévy, 2003, p. 997). The presence of irregular migrants is limited in time and restricted in space to specific areas; their personal space hovers between visibility and invisibility, pushed this way and that in response to the mass of constraints and the degree of tolerance shown by authorities, on the one hand, and to their ability to accommodate, assume risk and manage their fear, on the other.

In Spain, a renewed interest in hiring domestic servants (known as *interna*) about ten years ago encouraged international female migration, since it is estimated that 80 to 90% of domestic servants are foreigners, most of whom do not have a job contract (Martinez Buján, 2012). This has led to domestic service becoming a stepping stone in countries of destination; over time, they are lodged by their employer and thus, by savings on transport and housing costs, are able to pay off debts incurred for their voyage more quickly or send more in remittances to their country of origin. For new migrants who have not yet learnt to decipher the social codes in the country of destination, it also allows them to minimize contact with the public and remain invisible most of the week: thus they do not suffer from fear of being visible in the public arena and are relatively safe from being arrested.

In interviews, many migrants have spoken of their fearful desire for invisibility and, at the same time, demonstrated a very strong incorporation of representations of boundaries due to their status. This fear can be expressed in terms of a need for discretion and restricted movements: they avoid going to city centres where checks are more likely or prefer to go out after dark, and then only to private homes, during major public holidays when spot checks are more likely. Based on 15 months of research (late 2010 to mid-2011), a civil society report presented many stories about migrants and their fear of participating in evening activities or of entering immigrant neighbourhoods, metro stations and public areas: for example, they take fewer visits to the park with children or prefer taxis to public transport for returning home from work, despite the additional cost (Brigadas vecinales de observacion de derechos humanos 2011).

Tighter controls, in particular racial profiling during identity checks and more identity checks unrelated to criminal acts, have sparked a large number of protests by associations, collectives,[26] police unions and international organizations. In Madrid, there have been repeated raids in Plaza Eliptica, the square where immigrants can be hired as day labourers (as often as several times a week in April 2010 (MUGAK, 2011)), in central districts with a high foreign population (Tirso de Molina, Lavapiés, Vallecas, Embajadores, Atocha, Estrecho, Tetuan, Cuatro

[26] In March 2010, a collective of 141 associations for the defence of human and migrant rights presented a petition to the Public Defender and the Ministry of Interior denouncing the unfairness of provisions in Circular 1/2010.

Caminos, Usera), at metro station exits, on city buses, and in parks and public places (Inmigrapenal, 2010; MUGAK, 2011; CARITAS, 2010; Brigadas vecinales de observacion de derechos humanos, 2011; Ferrocarril Clandestino, 2010). Checks are also carried out near government buildings, "*places where foreigners go to obtain permits (consulates, embassies), associations and social centres, places they visit as part of their normal daily life (telephone centres, health centres, public transport stations, leisure areas..)*" (Inmigrapenal, 2010). Others spoke of arrests close to schools where parents come to pick up their children and even in association centres.[27]

Manoeuvres and diversions: new forms of mobility

In the United States, checking identities when no offences or infractions have taken place is a new development and is limited to certain states that have introduced tougher anti-immigration legislation, such as Arizona. Nevertheless, implementation of specific programmes, including Arizona's 287g and *Secure Communities* mentioned earlier, worries associations for the defence of migrants. These associations have reported that undocumented migrants are more prey to fear and have withdrawn into the privacy of their homes. This, in part, explains why some States decided to pull out of these programmes in 2011. The ways that people handle their clandestinity on a daily basis differ markedly from one city or state to another in response to the arsenal of controls available under the legislation. The geography of flight or manoeuvre coexists with a geography of fear, as we shall see in the following examples.

In October 2011, Alabama, where the largest foreign minority is Hispanic, passed one of the most restrictive anti-immigration acts in the country (HB56), which requires (a first for the United States) that schools register the legal status of their pupil. Immediately the press claimed that, in the weeks following its enactment, the number of Hispanics in schools dropped by 5%.[28] Was this a temporary retreat brought on by fear or the beginning of an exodus of residents, as could be deduced from reports that rented lodgings were abandoned in the Hispanic neighbourhoods in the State over the same period? In cities with repressive legislation, the closure of Hispanic stores seems to

27 Caritas operates 68 drop-in centres in Spain, of which 54 participated in the MUGAK report (2010). According to this report, more than a third of the 400,000 persons who visited Caritas' centres were undocumented migrants (150,000) and many of these were unable to renew their work permit because of rising unemployment in Spain (higher than 30% for foreigners). Caritas estimates there are 500,000 illegal workers in Spain.

28 Cf. Editorial and articles in the *New York Times*. "Alabama's shame", 3 October 2011.

indicate that they are fleeing to other more welcoming cities and states in order to avoid new legal constraints on their activities.

The Federal *Real ID Act* 2005 aims to improve the credibility of proofs of identity and, in particular, making the issue of driving licences dependent on proof of legal residence. Irregular migrants thus find themselves unable to drive legally in the United States. In 2012, only two States (New Mexico and Washington) do not require proof of legal residency before issuing a driving licence (Waslin, 2012). This measure has dramatic consequences for undocumented immigrants in a country where the car is king, since without one they are unable to move around and in particular go to work. Interviews with Latin American migrants and consulate personnel in 2011 in the Washington DC metropolitan area, which covers not only the city of Washington, but also parts of Virginia (where most Latin American migrants live) and Maryland, confirm the results of several studies (Capps *et al.*, 2011): although Maryland has more flexible laws which allow residents of Virginia (reputed to have much stricter migration policies) to obtain a driving licence but, since 2010, when it became obligatory to produce proof of local residence, more and more people are resorting to cross the country in the hope of obtaining a licence in the State of Washington (more than 4500 km away!), while many continue to drive, but without a licence. Fear of arrest for minor infractions (non-use of seat belts, lack of indicators, etc.) which, thanks to the new regulations allowing access to federal files, can now lead to an expulsion order, has not only forced migrants to become more prudent but has also the risks (not reporting or fleeing the scene of accidents, etc.) and generally to reduce their use of cars for anything but absolutely essential journeys. Thus, changes in American legislation can have a profound impact on the mobility, residence and daily life of undocumented migrants in a country where their rights differ considerably from one area to another and generate new forms of mobility that reinforce local immobility.

Protection and dependence in ethnic districts

The obligation to be discreet does not mean that irregular migrants disappear completely from local society. The high number of ethnic neighbourhoods, reflecting as much the many way of integrating as of segregating, indicates migrants' strong dependence on ethnic networks based on areas where they first settled and often on their country of origin and their reliance on them for assistance on their arrival and in daily life. These ethnic links operate at various levels: by opening doors into the labour market, when a member of the group provides information on jobs for clandestine workers, or by passing on tips on the housing market. Faced with the difficulties of obtaining lodgings because of a lack of guarantees, deposits or papers, irregular migrants often move in with relatives, or hire a bed or a room in an apartment rented to a legal resident prepared to sub-let to their clandestine compatriots. Given that there are high levels of promiscuity in cheap apartments in large cities and towns, this tendency to share is very common in ethnic communities: migrants will prefer to share a room with someone who speaks their language and shares their cultural origins and references. Ethnic districts also offer opportunities for maintaining links with the country of origin: agencies for sending remittances and parcels and cybercafés with low rates for telephoning relatives left behind, all also places where migrants can find ethnic newspapers and classified ads for jobs and lodgings. Together with restaurants, bars, and food stores that offer familiar products from home, they are also places where migrants can learn about the new environment in which they find themselves.

To a certain extent, undocumented migrants feel themselves protected by the group and these institutions, as can be seen in research on ethnic enclaves (Wilson and Portes, 1980; Waldinger, 1993). Hierarchies within these groups (the least stable migrants depending on those with greater experience, irregular migrants on legal residents) can also create occasions for disputes and opportunities for exploitation and conflict. Omar, a young man of 18 in search of adventure in Europe, talked about his clandestine stay in Sweden and described not only the scams organized within Stockholm's Bolivian community where miserable flats are available at prohibitive prices to people with no other options, but also the generous offers of help and tips: how to claim asylum in order to be able to stay longer or where to find free classes in Swedish and soup kitchens, etc. (Montecinos Escalier, 2008). These neighbourhoods provide protection and a certain amount of familiar interaction, where irregular migrants may be dependent on others, where information from home is shared, and where it is possible to put down roots in the new country. They are, at the same time, dangerous places in which police controls are organized, under the tougher policies described above, and therefore potentially high-risk zones for undocumented migrants.

Comparison of Spain and the United States: Perspectives

As they learn to negotiate the rules in their daily life, undocumented migrants begin to move around with a certain degree of confidence, but their mobility is limited because of tougher policies and greater controls on migration. Insecurity and fear are a daily reality throughout Spain, and especially in major cities and neighbourhoods with high migrant populations, whereas in the United States, the situation is much less uneven because the wide range of existing laws and forms of control which vary from one county and one state to another and thereby create zones with high levels of tolerance and fewer police controls in more protective states and counties.

In the United States, because of the very high number of undocumented migrants, their presence has become almost 'banal': they are more or less socially acceptable and are not considered to be criminals *per se*, despite tighter police controls. Spain, with fewer irregular migrants, tends to stigmatize them and make them feel more unstable in society, although they are 'irregular' for shorter periods since regularization is possible after three years of residence. The tendency for migrants to remain irregular for long periods of time in the United States is also a key element in their social acceptability. The DREAM Act[29] in particular has sparked many debates, though Barack Obama now announced, in June 2012, that it will apply to certain categories of migrants. These debates turned the spotlight on the situation of many young undocumented adults who migrated as children and grew up in the United States: many now have university degrees but are blocked from entering the job market because of their long-term irregular status, sometimes after 20 years in America.[30] This type of long-term clandestinity is specific to the United States and explains why such a very wide range of undocumented migrants is not found in Spain, since most undocumented migrants are recent arrivals and able to find unskilled jobs either legally or in a thriving black market. Living without papers in Spain is more likely to be a temporary phase that will end after a few years whereas, in the United States, illegal migration is almost an apprenticeship in clandestine life which, despite a certain tacit tolerance, may extend indefinitely into the future. Tighter migration policies probably have a greater impact on recent migrants in Spain than in the United States where they merely contribute to making migrant populations more vulnerable and more prone to crime. However, migrants in the US tend to become better socially integrated in the long term and find, despite their irregular status, greater acceptance and tolerance of their presence despite the repressive legislation.

---

[29] The *Development, Relief and Education for Alien Minors Act* has been under discussion for 10 years: it would allow conditional regularization for young foreigners who have lived and studied in the United States (Batalova and McHugh, 2010).

[30] José Antonio Vargas, a young Filipino journalist recently awarded the Pulitzer Prize, is a spokesperson for this generation of undocumented graduates, after making his coming-out as an 'undocumented' migrant, in the pro-DREAM Act campaign. He has written articles describing the daily struggle of migrants to obtain their rights while living in persistent irregularity.

Conclusion

For migrants, their status has a strong impact on their lives. It affects the conditions of their departure and crossing of frontiers and the development of family strategies and, at the same time, it creates discrimination in their daily life in countries of destination. By influencing how and where they can obtain jobs and housing, their irregular status also restricts their mobility and their visibility is affected by the implementation of migration control policies. The tightening of these policies puts barriers around the daily environment of irregular migrants and, in the United States, the wide range of legislative measures for controlling migrants and the disparities in the way states and municipalities manage their implementation also affect their mobility, either by reinforcing their immobility or by forcing them to move to less controlled areas. As migration becomes increasingly global, we can learn much about the movement of diasporas at their various levels, by examining the changes in status, opportunities and constraints within areas of migration. If, on the one hand, being irregular reduces one's ability to travel across national borders and leads to greater immobility, on the other, this situation can encourage them to develop tactics of adaptation to such constraints and become more mobile, which is paradoxical in a situation of denied mobility (frontiers closed to 'illegal' migrants) and of a form of rootlessness, a *deportspora*.[31]

Is regularization the key to integration? Often described as the '*key to the world*' or a '*180° reversal*',[32] it obviously opens the door to their individual rights as legitimate citizens and allows them to give up tactics of invisibility, manoeuvres and flight and simply affirm their presence. Nevertheless, some migrants are disappointed by their hopes when regularization and the acquisition of papers does not '*changed their lives much*' and recall that obtaining the right to live there, even if it is crucial, is only a first step and does not resolve all their economic and family problems or protect them from discrimination.

References

Agier M. 2008. *Gérer les indésirables: du camp de réfugiés au gouvernement humanitaire.* Paris: Flammarion, Collection Bibliothèque des Savoirs.

Agier M. (ed.) 2012. *Politiques de l'exception – Réfugiés, sinistrés, sans-papiers.* Paris: Tetraèdre, le sujet dans la cité.

Baby-Collin V. and D. Mercier (eds.) 2012. *Le Mexique dans les migrations internationales.* Paris: Hommes et Migrations.

[31] A combination of *deported* and *diaspora*, N. De Genova, 2010, 26.

[32] Interviews with irregular and regular Latin American migrants, mainly from Andean countries: New York and Washington DC, 2011-2012; Madrid, 2008-2010.

Batalova J. and M. McHugh. 2010. *DREAM vs. Reality: An Analysis of Potential DREAM Act Beneficiaries.* Washington DC: Migration Policy Institute.

Bélouin S., A-L. Counilh, O. Bronnikova and S. Mekdjian. 2009. *Le visible et l'invisible dans le champ des études sur les migrations*. Paris: E Migrinter.

Bloemraad I. and E. De Graauw. 2012. "Immigrant integration and policy in the United States: a loosely stitched patchwork". *International Perspectives: Integration and Inclusion.* Montreal: McGill-Queen's University Press, Queen's Policy Studies Series. 205-232.

Brick K. 2011. *Regularizations in the European Union: The Contentious Policy Tool*. Washington DC: Migration Policy Institute.

Brigadas vecinales de observacion de derechos humanos. 2011. *Controles de identidad racistas en Madrid, Informe.* Madrid: Brigadas vecinales de observacion de derechos humanos.

Capps R., M. Rosenblum, C. Rodriguez and M. Chisthi. 2011. *Delegation and Divergence: A Study of 287(g) State and Local Immigration Enforcement*, Washington DC: Migration Policy Institute.

CARITAS, 2010. *La situación social de los inmigrantes acompañados por Caritas*. Madrid: CARITAS.

CERES y CCOO. 2004. *Inmigracion irregular en Cataluna. Analisis y propuestas.* Barcelona: CERES CCOO.

Certeau M. de, L. Giard, and P. Mayol. 1990 (1st ed.: 1980). *L'invention du quotidien, tome 1: Arts de faire.* Paris: Gallimard.

Cornelius W.A. 2005. "Controlling 'unwanted' immigration: Lessons from the United States, 1993-2004". *Journal of Ethnic and Migration Studies*, 31: 775-794.

Coutin S.B. 2007. *Nations of Emigrants: Shifting Boundaries of Citizenship in El Salvador and the United States*. New York: Cornell University Press.

Ferrocarril Clandestino. 2010. *Informe de investigación sobre controles identificaciones y detenciones.* http://brigadasvecinales-ddhh.blogspot.com/2010/01/objetivos-de-las-brigadas-vecinales-de.html (Accessed June 6, 2011).

De Genova N. 2010. *The Deportation Regime: Sovereignty, Space, and the Freedom of Movement*. Durham, NC: Duke University Press.

De Genova N.P. 2004. "The legal production of Mexican/ migrant 'illegality'". *Latino Studies*, 2: 60-185.

Gonzalez Enriquez C. 2009. *Undocumented Migration. Counting the Uncountable. Data and Trends Across Europe. Country Report. Spain.* Clandestino Project. Brussels: European Commission.

Gonzalez Enriquez C. 2010. "Spain: Irregular Immigration as a Rule", in Triandafyllidou A. (ed.), *Irregular Migration in Europe: Myths and Realities.* Farnham, UK: Ashgate Publishing

Hanson G. 2009. *The Economics and Policy of Illegal Immigration in the United States*. Washington: Migration Policy Institute.

Inmigrapenal. 2010. *Informe controles de identidad y detención de inmigrantes, prácticas illegales*, Madrid: Inmigrapenal. http://brigadasvecinales-ddhh.blogspot.com/2010/01/objetivos-de-las-brigadas-vecinales-de.html (Accessed June6, 2011).

Kerwin D.M., K. Brick and R. Kilberg. 2012. *Unauthorized Immigrants in the United States and Europe: The Use of Legalization / Regularization as a Policy Tool*. Washington DC: Migration Policy Institute.

Levinson A. 2005. *The Problem of Regularization of Unauthorized Migrants, Literature Survey and Country Case Studies*. Oxford: COMPAS.

Lussault M. and J. Lévy. 2003. *Dictionnaire de la géographie et de l'espace des sociétés*. Paris: Belin.

Martinez Buján R. 2012. "De sirvientes a enfermeras. Transformaciones en el servicio doméstico español". Conference Papers, S. Durin (ed.) *Dimensiones del emleo domestico latino-americano*. Monterry, Mexico, 16-18 May 2012.

Meissner D. and M. Rosenblum. 2009. *The Next Generation of E-verify, Getting Employment Verification Right*. Washington DC: Migration Policy Institute.

MIGREUROP and O. Clochard. 2009. *Atlas des migrants en Europe*. Paris: Colin.

Mittelstadt M., B. Speaker, D. Meissner and M. Chisthi. 2011. *Through the Prism of National Security: Major Immigration Policy and Program Changes in the Decade since 9/11*. Washington DC, Migration Policy Institute.

Montecinos Escalier O. 2008. *Ser un migrante en 'las Europas', manuel de sobrevivencia en el 'viejo mundo'*. La Paz: Julyo's.

MUGAK. 2011. *Los controles policiales en la vía pública dirigidos a la identificación y detención de las personas migrantes irregulares*. Centro de estudios y documentación sobre racismo y xenofóbia de SOS Racismo, SOS Arazakeria. http://brigadasvecinales-ddhh.blogspot.com/2010/01/objetivos-de-las-brigadas-vecinales-de.html (Accessed June 6, 2011).

Myers D. 2007. *Immigrants and Boomers: Forging a New Social Contract for the Future of America*. New York: Russell Sage Foundation.

Revue du MAUSS. 2004. *De la reconnaissance. Don, identité, estime de soi*. Paris: La Découverte.

Riosmena F. and D.S. Massey. 2012. "Pathways to El Norte: Origins, destinations, and characteristics of Mexican migrants to the United States". *International Migration Review*, 46(1):3-36.

Simon G. 2008. *La planète migratoire dans la mondialisation*. Paris, Colin.

Sumption M. 2011. *Policies to Curb Unauthorized Employment*. Washington DC: Migration Policy Institute.

Triandafyllidou A. 2010. *Irregular Migration in Europe: Myths and Realities*. Farnham, UK: Ashgate Publishing.

Urry J. 2005. *Sociologie des mobilités*. Paris: Colin, Collection U sociologie.

Voirol O. 2005. "Présentation : Visibilité et invisibilité, une introduction". *Réseaux* 129-130(1): 9-36.

Waldinger R. 1993. "Le débat sur l'enclave ethnique : revue critique". *Revue Européenne des Migrations Internationales* 9(2):15-29.

Waslin M. 2012. "Driving while immigrant: driver's license policy and immigration enforcement". *Issues of Immigration in Criminal Justice*. John Jay College of Criminal Justice, New York City.

Waslin M. 2011. *The Secure Communities Program: Unanswered Questions and Continuing Concerns*. Washington DC: Immigration Policy Center.

Wilson L.K. and A. Portes. 1980. "Immigrant enclaves: An analysis of the labor market experiences of Cubans in Miami". *American Journal of Sociology* 86(2): 295-319.

Part IV

Circulation, Frontier and Memory of Migrations

Introduction

Stéphane MOURLANE, Pierre SINTÈS

The question of the memory of migration implicitly focuses on the object-frontier and treatment of this phenomenon can provide useful information on changes over recent decades. Until now, according to studies carried out mostly in their country of installation, migrants tended rather to be destined to a form of programmed forgetfulness as a result of their long-term residence in a new country. This is the conclusion that could be drawn, for example, from the writings of Abdelmayek Sayad who analyzed "the three ages of Algerian migration in France" in 1977. Sayad saw migration as a series of phases that inevitably led migrants towards a more or less successful insertion into the destination country and the disappearance of links with their land of origin. This kind of approach is the source of well-established typologies and classifications and is based on the reasons behind the decision to change country (pioneers, workers, families). All implicitly suggest a teleological approach towards migratory mobility, in that actors seem to pass inexorably, depending on the length of their stay, from the status of 'new arrival' to that of a fully established 'immigrant' with family. Such visions were appropriate at the time they were put forward. Indeed, the 1970s and 1980s saw the deliberate closure of European frontiers, even though integration seemed to be obligatory for foreigners seeking to contribute to the Old Continent's economic reconstruction. As Dominique Schnapper (1992) has stated, zero immigration and closed frontiers were, at the time, a corollary of the expectation that foreigners would assimilate, in France as elsewhere in Europe. In a very conformist way, the paradigms that must be taken into account in migration studies present migrants either as immigrants in the 'process of integrating' (often with difficulty) or as persons absent from their country of origin while contributing to its development as emigrants.

Counter-flows (whether material or immaterial) have only gradually become essential and legitimate objects for study, thus highlighting relationships with the country of origin (Simon, 1979). They can be expressed as journeys between countries of installation and of departure, regular remittances to families, and multiple contacts made possible and more visible by technological progress. Finally, this sometimes imaginary relationship with this country seemed to contribute to developing strong links to migrant associations, as part of diasporas rediscovered in the late 1990s. The importance of links with the country of departure and that of destination has provided new themes for migration studies, since these links have emerged as structural elements in the lives of migrants.

These considerations make it possible to pay greater attention to social phenomena that focus on issues relating to identity and memories with regard to migration. This new perspective could even be widened in order to attribute the creation of specific social worlds, a true 'culture of mobility', to migration (Clifford, 1992). At the same time, this type of initiative can contribute to a partial reconsideration of migration in terms of phases, since relations with the home country change considerably over time. Revisiting the theme of common origins can be seen clearly in what we could call a more noticeable 'age of return' than in the past. It also interacts with two other phenomena that have attracted the attention of researchers.

First, a '*retour sur soi*', a reflection on personal experience and the individual's migrant roots as a way of *rehabilitating the memory of migration* in order to give value to a history that is often little-known or neglected by official discourse. In such circumstances, these moves contribute to research into the complex relationship between history and memory that certainly needs to be analyzed without, nevertheless, systematically opposing the two concepts (Ricœur, 2000) and, in particular, it facilitates the emergence of new themes for historical studies. Mobilizing memory for groups of migrants, from a trans-generational perspective, can also lead to the creation of commemorative activities: ceremonies, artistic or documentary works, protection and valorization of heritage sites are all ways of reappropriating a forgotten or sacrificed past, often with the collective aim of developing a national fiction.

The desire to rebuild a migratory past can also be fulfilled by creating *new forms of geographic mobility* which, as a counter movement to the initial migration, encourages migrants (or their descendants) to return to their past and their country of origin, either through regular journeys (holidays or dual purpose) or for more unusual voyages (pilgrimages of identity, roots tourism).

As during the previous period, during the 1970s and 1980s, the emergence of these new fields of research can no longer be separated from their context, due to the impact of a dual dynamic that sees the questions of migration and of links with the past become social, cultural and political issues of the highest order. By the 2000s, the political debate on the identity of national groups and the manner in which migrants seem to defy or undermine this identity were at the centre of discussions on migration themes that focused on a return to the theory of civilizations. It is no accident that, during this period, interpretation of the term 'diaspora' widened considerably, as an indication of a different approach and of the perception of migration memory. Memory may be well-established, even tenacious, and is not easily lost by migrants in their country of settlement, since they are never completely settled, as implied by the new paradigm now being imposed on migration studies (that of circulation), but also since they may be going through a process of cultural hybridization – and therefore not of forgetfulness or transition, as suggested by Sayad. This new attitude towards migration suggests a more radical reconsideration of the figure of migrants: as members of a transnational community, for whom history offers a way of explaining their practices and acts – and for several generations, for as long as memory remains and is active (or activated).

It must be noted that attention to the memory of migration is itself part of a more general movement during the period, what Marcel Gauchet (1998) calls 'the age of identities' according to which various groups will evoke the past, sometimes in order to assert themselves or legitimize their presence, and, for some, give shape to a collection of 'waves of surging memories' (Prochasson, 2008). Having entered our everyday language as a way of reviving a moral or political obligation to remember, the 'duty of memory' has become commonplace in the media, statements by politicians, and speeches by religious and civil society leaders, to the point where it seems to become a 'new civil religion' (Bensoussan, 1998). The abundance of initiatives and the emotional burden that they generate sometimes create 'competition' amongst various types of memory in a more general social climate of a 'war of memories' (Blanchard and Veyrat-Masson, 2008) that only few countries are able to avoid, especially in the Mediterranean Basin. Fragmented representations of the past, which mostly occur during these memorial events, inevitably raise the question of co-habitation. The proclamation of supra-national and inclusive identities, European or Mediterranean for example, often together with mythical overtones and the eternal perspective of integration, goes hand-in-hand paradoxically with the revival of demands for memorials based on the difference in which migrants themselves are now active participants.

The four contributions in this section seek therefore to shed light on the complexity of the memorial process and its relationship with migrations and frontiers, which are seen as markers of both territory and identity. Despite the variety of cases studied, in terms of time and space, all four articles highlight the lines of convergence that traverse these situations. Stéphane Mourlane and Céline Regnard, together with Karima Dirèche, remind us that forgetting usually precedes remembering.

For Italian migrants in France or Algerian Jews, forgetfulness is linked to expectations of 'francization' or assimilation, as required by Republican ideology. Taking as his example 19th century Italians settled in the Orient, Fabrice Jesné shows that the memory of migration has few resources when national identity, which in his example is still in gestation, lacks the solid cement of national construction. The authors have, for the most part, focused on methods for reactivating memory, as discussed by Pierre Sintès in his article on Greece, necessarily rooted in – often severely restricted – social and territorial realities. Stéphane Mourlane and Céline Regnard insist on the different levers and vectors of Italianness at work in the resurgence of memories of trans-Alpine migration in France. The role of associations and the media in launching numerous cultural events is particularly important, as shown in the case of Algerian Jews. Karima Dirèche sees 'identity constructions of variable geometry' in these reminders to remember, an approach that Pierre Sintès confirms when he emphasizes the 'polyphony of identities' at the local level. All these contributions focus on the relationship between history and memory, notably from the perspective of the ambivalent role played by historical research in the process of memory as agents for distancing or bringing closer together migrants and their country of destination. Of course, there are just a few ideas in a vast field of research that is still very much in its infancy.

References

Bensoussan G. 1998. *Auschwitz en héritage? D'un bon usage de la mémoire*. Paris: Mille et une nuits.

Blanchard P. and I. Veyrat-Masson (ed.), *La Guerre de mémoires. La France et son histoire*. Paris: La Découverte.

Clifford J. 1992. "Travelling cultures", *Cultural Studies*. New York-London: Routledge: 96-116.

Gauchet M. 1998. *La Religion dans la démocratie: parcours de la laïcité*. Paris: Gallimard.

Prochasson C. 2008. *L'Empire des émotions: Les Historiens dans la mêlée*. Paris: Démopolis.

Ricœur P. 2000. *La mémoire, l'histoire, l'oubli*. Paris: Le Seuil.

Sayad A. 1977. "Les trois 'âges' de l'émigration algérienne en France", *Actes de la recherche en sciences sociales*, 15(15): 59-79.

Schnapper D. 1992. *L'Europe des immigrés*. Paris: François Bourrin.

Simon G. 1979. *L'espace des travailleurs tunisiens en France, structures et fonctionnement d'un champ migratoire international*. PhD thesis published by the University of Poitiers.

Migration and Nation

The Situation of Italians in the Eastern Mediterranean in the Early 19th Century

Fabrice JESNÉ

According to Todorov (1995), memory "gives priority to the immaterial world of psychic experiences" but where do we find references to memory in the history of migration between the Italian peninsula and the Eastern Mediterranean during the 19th century? There have been Italian communities in the Orient since the Middle Ages, but the fall of the Ottoman cosmopolitan world renders the task of defining the 'place of memory' (Nora, 1984, XVII) for these communities much more difficult (Schmitt, 2007). However, with the gradual dissemination of the national identity during the 19th century, it is possible, in the Eastern Mediterranean as elsewhere, to see attempts at 'inventing traditions' (Hobsbawm and Ranger, 1983) which we could see as ways of creating memory or, *ad minima*, of contributing to their emergence. Since the 1970s, historians and anthropologists have denaturalized the concept of nation, showing how it emerged towards the end of the modern era, at the same time as patrimonial sovereignty and other forms of solidarity linked to the Old Regime began to decline. First used in Western Europe, the concept of 'nation' spread to the rest of the world. Although initial studies relating to the invention of national identities insisted on a cultural construction of the concept, later research has focused on how it was adopted by the masses and the importance of their acceptance of a proposed 'nationhood' via a reinvention, for the benefit of the nation, of older forms of material and symbolic relationships, (Andersen, 1983; Gellner, 1983; Hobsbawm, 1990; Thiesse, 1999).

With regard to these symbolic links, we will examine the possible forms of memory of 'Italy' by analyzing the plurality and malleability of the national identity-in-construction based on a wide range of traditional activities. The context of an "Italy outside Italy" (Pambianco, 1926), and more precisely in the Orient, offers a suitable basis for such an analysis, in that memory of Italy offers opportunities for studying a very ancient history.

Italian merchants set up trading posts along the Southern and Eastern coasts of the Mediterranean during the Middle Ages (Balard, 1978; Basso, 1994; Dursteler, 2006). The presence of Christians in Muslim countries led to the establishment of an extra-territorial regime and consequently to a consular administration. Italian merchants, and later other Christian traders, obtained the right to live in specific districts under the jurisdiction of a consul. This practice spread throughout the regions where the Ottoman Empire granted 'capitulations' to various Western powers, under which the consul assumed not only the traditional role of community leader, but also that of representative of the community's European state of origin (Van den Boogert, 2005).

In this situation, the demographic and cultural heritage based on Italian culture was gradually diluted over the centuries to such an extent that, by the 19th century, descendants of Christian merchants in the East were known as 'Levantines' by Western Europeans; many of these Levantines preserved a family tradition of their European origins, which were often Italian, of course, but also French, English, Spanish and other, including Oriental, origins, since Europeans often married into local communities, notably Greek, Jewish, and Armenian (Schmitt, 2007).[1] Also, the *lingua franca*, a language similar to Italian, was spoken throughout the Mediterranean, as a neutral or 'no man's language' without a specific identity (Dakhlia, 2008).

Even though in today's Italy it is often tempting to refer back to its long history from Imperial Rome, via Dante's era and other major periods, to the present day,[2] in the early 19th century, Oriental Italians did not see themselves as Italian in the sense that we understand it today, any more than the vast majority of people living in Italy. If we accept Eric Hobsbawm's theory (1990), only part of the European elite recognized, at the end of the modern era, that they were part of a 'pre-national' community linked to dynastic past, and the masses usually saw themselves as 'peasants', in the absence of any national term for self-designation. In Italy, we must remember that, at the time of its unification (1861), only a very small percentage of its inhabitants spoke Italian. However, Oriental Italians were not peasants and perhaps, on the contrary, they were already developing a specific identity that became a precursor for affirming a national identity. Various Western powers had succeeded, by signing the Ottoman capitulations, in setting up extra-territorial regimes for their citizens in the Orient: these regimes were

1 It is difficult to give an exact estimate the number of Levantines, since it was in constant movement. The figure could be several tens of thousands.

2 "*Uniti da 150 anni, italiani da sempre*" ("United for 150 years, but always Italian") was the motto of the official celebrations for the 150th anniversary of Italian unity on 16 March 2011.

administered by a consul, who was not only the head of the community, but also magistrate, registrar, notary, and provost. Of course, no Italian communities, *per se*, did not exist but other groups had established 'colonies' of Sardinians, Neapolitans, Tuscans, even of Papal States residents, as well as subjects of the Austrian Emperor from Italian regions who formed part of the Italian cultural community established in Oriental ports.

Claiming to be Italian can thus be seen as part of social, commercial and administrative interaction (Bourdieu, 1980).[3] At the beginning of the 19th century, the Kingdom of Sardinia claimed sovereignty over Genoese colonies in the East. The Congress of Vienna had attributed all the former Republic of Genoa's territories, which had been taken by France between 1797 and 1814, to the ruling house of Savoy. Sardinia was among the group of states to which the task of monitoring France was designated; for the Great Powers, it was appropriate that its capital, Turin, be strengthened by access to the Mediterranean and by trade. Sardinia began to modernize its administrative and economic structures which involved, amongst other initiatives, official support for external trade. As a result, priority was given to controlling Genoa's former trading networks and establishing an extensive network of consuls.[4]

Like the other Powers that benefited from the capitulations, Sardinia could grant its 'protection', i.e. its extraterritoriality regime, to its own citizens and certain subjects of the Ottoman Empire and other states linked to Sardinia through its capitulation. These protégés were generally employees of Sardinia's consular administration (court clerks, secretaries and dragomans) and, for the most part, were members of minority non-Muslim populations who specialized in assisting relations between Muslims and Westerners (Heyberger and Verdeil, 2009): Jews, Greeks, Armenians, known as 'Levantines', and they often had at least some Italian antecedents. However, even if they could claim Italian blood, it was not always recognized by the Italian authorities, which preferred to offer their protection in terms of their consular services. Nevertheless,

---

3 Endogenous and exogenous attribution of origins is subject, of course, to the "specific argument in society, this 'reality' which is the cornerstone of the permanent struggle to *define* 'reality'".

4 "[The Sardinian Legation in] Constantinople had jurisdiction over the consulates of the Dardanelles, Varna, Salonica, Smyrna, Adrianopolis, Gallipoli, Tekirdag, Tenedos, Trebizond; [the consulate at] Smyrna was responsible for [the delegations] of Ayvali, Candia, Chania, Mytilene, Rhodes, Chios, Limassol; [the consulate at] Jerusalem: Gaza and Jaffa; [the consulate at] Beirut: Acre, Alep, Alexandretta, Haifa, Lattakieh, Ramallah, Saida, Tarsus, Tripoli, Sour; [the consulate at] Alexandria: the consulate in Cairo and the proconsulate in Damietta" (Mola di Nomaglio and Sandri-Giachono, 2007, 320).

descendants of Genoese traders benefitted from special treatment from the Sardinian government:

> There should be no confusion between Sardinians and people who, although born and living abroad, were originally from the kingdom and have always been considered and treated as royal subjects. This is the case, in the Levant and the Island of Chio, for numerous families of Genoese origin. These royal subjects will treated in the same way as in the past.[5]

In this case, it is principally a question of guaranteeing administrative continuity between the Republic of Genoa and the Kingdom of Sardinia which, in the circumstances, was very interested in maintaining and extending the commercial potential of its new administrations in the Orient. Nevertheless, origins are the basis for family memories and the first prerequisite for establishing one's nationality. We have seen also that the consuls exercised their jurisdiction, in the Orient as elsewhere, with reference to 'treaties' (*trattati*) and 'customs' (*usi*):[6] in many cases, tradition had force of law and thus, for 'native' Genoese, this was the origin of their nationality: the now smaller Genoese homeland was absorbed into the larger Sardinian identity, at least until an Italian nationality came into being.

In the same way, the Grand Duchy of Tuscany had, among its subjects in the Orient, a large number of Portuguese Jews whose families had been living and trading in Livorno and the East. One of these was Isacco Saul Sonnino, father of the famous Sidney, a banker in Alexandria. Through his mother, Sidney Sonnino also was related to the Castelnuovo, another Livornese Jewish family which was also established in Tunis: in these families, the first step was usually to go and study in Tuscany before "*being returning to the local Livorno communities*" in the Orient (Levy, 1999, 99-100). Paradoxically, it was the abolishment of long-established grand-ducal franchises by the newly united Italy that forced many of these families to leave Livorno, despite the new regime's support for the Jewish community (Milano, 1949, 186-189). At the time of unification, Sardinia's annexation of the pre-unity states that were to become Italy led to a fusion of consular networks in the Orient or rather their replacement by a single Sardinian network and its agents who would now represent all the Italian communities. In the

5 "*Coi protetti sardi non sono da confondersi coloro che, quantunque nati e domiciliati in paese estero, sono però oriundi dei R.[i] Stati e vennero costantamente considerati e trattati come regii sudditi, quali sono in Levante, nell'Isola di Scio, molte famiglie di origine Genovese. Questi regii sudditi di origine continueranno ad essere trattati come per lo passato*", in Ministero degli Affari esteri (MAE), *Ordinamento consolare. Istruzioni ai consoli di S. M. il Re di Sardegna*, Turin, Stamperia Reale, 1859, MAE, 1859, art. 23.

6 *Ibid.*, art. 62.

same way that it had sought to highlight the demographic and commercial heritage of Genoa, Sardinia set about, with the same enthusiasm, its campaign to recuperate Tuscany's network which was quite different from the peoples of Naples and the Papal States: we will return to this point below (Jesné, 2009: 86-89).

Thus, in the Orient, there were Levantine populations with partially Italian origins and others from so-called Italian states: Sardinia, Naples, Papal States, Tuscany, amongst others. Even during periods of mass migrations across the oceans, migrants have never lost contact with their homeland and continued to make frequent voyages there and back (Franzina, 2006, 38). In fact, these voyages nurtured and strengthened their links with a smaller 'homeland' – their village, town, or region – rather than with the larger '*patria*', in particular for the pre-unity states. In the 1850s, there were 'papal subjects', known as 'Romagnols' living in Athens, because Greece was the closest foreign country after Austrian Dalmatia. Similarly, Neapolitan fishermen were in fact Sicilians from Trapani.[7] Fishermen were particularly mobile, and it is difficult to classify this mobility as 'migration', a term too explicit for such a complex reality. We must therefore establish some characteristics of 'Italian' mobility in the Orient during the early 19th century, a theme somewhat neglected by historical research which has focused more on contemporary trans-oceanic migrations (Audenino and Tiribassi, 2008; Bevilacqua *et al.*, 2001; Corti and Sanfilippo, 2009; Franzina, 1994; Gabaccia, 2000).

Italians' reasons for leaving the Peninsula were first and foremost commercial. Alongside the Levantine families who traded with Italian States, merchants from the Italian mainland were also present in the major Ottoman trading posts. Very often, they did not have a permanent residence, but lived there for a number of months or years. For example, Adriano Lemmi, son of a rich merchant from Livorno, established his business first in his home town and in Naples. He then began to develop his business around the Mediterranean, in Marseilles and Malta, where he took a wife, and then in Egypt and Constantinople, where he lived for a few years in the mid-1840s and again in the 1850s. After acquiring property in the Ottoman capital, he settled in Italy in 1860. Even then, his voyages took him to England and the United States, where he was able for a while to claim American nationality.[8] Such an important well-travelled merchant was not the exception and we find many other, more anonymous, personalities in the consular archives, including sailors.

7 *Archivio Storico del Ministero degli Affari Esteri* (ASMAE), *Scritture Sarde* (SS), *busta* 250 (*Consolato sardo ad Atene*).

8 Lemmi, Adriano, *ad nomen in Dizionario Biografico degli Italiani*, Vol. 64.

Fishing and commerce brought teams of Italians, mainly from Sicily, Apulia and the Marches, to Greece and the Ottoman Empire.[9] Because all cargos had to be registered and taxed, there are references to their passage in consular statistics and the correspondence of consuls, particularly when referring to the many disputes involving ship-owners who constantly sought to avoid registration with health and fiscal authorities. Smuggling was common, especially if we take into account the fact that Mediterranean ships carried very small tonnages and therefore required only small ports for unloading: 'informal ports' as these are called by historians – often no more than a creek – were sometimes the norm for smuggling (Carrino and Salvemini, 2006). It was therefore very difficult to calculate how many times Italian subjects travelled to the Orient, although a systematic analysis of consular archives would certainly allow us to establish some idea of orders of magnitude.[10]

Labour migration represents a second characteristic of Italian mobility in the Eastern Mediterranean. Here, historians have focused more on emigration to the American continent and Western Europe during the late 19th and the 20th centuries, neglecting somewhat both migrations to the Eastern Mediterranean and earlier waves during the early 19th century. And yet, Italians first travelled to the Orient to find work during this period. This type of mobility was often not final and, for many, just seasonal migration.[11] At that time, it was very easy to travel from Italy to the Balkans by crossing the Adriatic and Ionian Seas. The Turkish Albanian coast was not very open to trade, but it was easier to set up short and regular voyages from Apulia; groups of Italian workers would make the trip in order to hire themselves out for work (Jesné, 2009). This was, in fact, a form of exported migration that was commonplace on the Italian peninsula: mountain villagers would go down to the plain in summer, often grouping together into brigades led by a 'corporal', to bring in the harvests. In the Balkans, this was known as '*gurbet*' (Petrov, 2008). Better qualified than local workers, Italians arrived in the Orient at a time when this region was just beginning to modernize its industry and lacked skilled labour. In Turkey's European territories, the silk industry lacked solid infrastructure and was interest-

9 ASMAE, SS, *buste* 250 (*Consolato sardo ad Atene*), 253 (*Consolato sardo a Corfù*), 262 (*Consolato sardo a Sira*).

10 This was one of the objectives for our Cosoli collective research project at the École française de Rome.

11 Many peasants left the Northern Apennines, in the mid-19th century, for places from London to Moscow and Marseilles to New York and Buenos Aires and, for the period that interests us, Istanbul, Thessalonica, Smyrna, Beirut and Alexandria. Most became cooks, waiters, ink sellers, organists, animal trainers, labourers, etc. (Zucchi, 1992, 23).

ed in hiring Italian workers and, in particular, Lombards who could handle every phase of silk production, whether simple or technical (Zucchi, 1992).

There are also references in the consular archives to a more unexpected form of emigration mainly from Southern Italy: Italian travelling artists have been studied in Western Europe (Zucchi, 1992) as 'proto-emigrants', the avant-garde of more substantial migratory movements (Franzina, 2006, 43); we find these '*girovaghi*' in the Ottoman Empire and some are seen, by the consular agents responsible for monitoring their presence, as persons of low morality.[12] It is true that, throughout Europe, the end of the 19th and early 20th centuries were the scene of socio-economic transformations which produced massive unemployment in rural areas. As a result, huge numbers of people became wandering vagabonds and were thought to be social troublemakers, a threat epitomized by the travelling musician (Zucchi, 1992: 7). Often banding together in small or family troupes, travelling musicians, many of whom came from the Kingdom of Naples, were met with suspicion by authorities and local populations. Subjected to close supervision by local officers, their poverty often led them to commit petty crimes that their mobility facilitated but which suspect them of subversive political activities, since vagabonds were thought to spread revolutionary ideas, even if this image was more fantasy than real (Genoino, 1989). When these musicians crossed the kingdom's frontiers, it was up to the consuls to examine their passports, check that they carried out their military service, and ensure their social and political morals.

Once musicians crossed the kingdom's borders, the consuls were responsible for checking their credentials and ensuring that they carried out their military service and that their morals, socially and politically, were sound.

Following Italian unification, Sardinian agents were responsible for merging the various Italian consular networks together and often complained about the necessity of supervising poor Southern Italians groups that wanted to mix with the local population but also with delinquent bands. Although this attitude echoed other prejudices that Italians experienced during the great migrations of the late 19th and early 20th centuries, it should be noted that this was how Italian consults judged their own compatriots (*connazionali*). Northern Italians came face to face, in the Orient as in their homeland, with their poor and lawless counterparts from the South (Moe, 1992: 64).

12 *Archivio di Stato di Napoli* (ASN), MAE, *buste* 2551 (*Consolato a Corfù*) and 2669 (*Consolati in Grecia*).

Committing ordinary crimes can often led their perpetrators to political crimes and we can suggest a last form of mobility linking migration and nation, political exile. The Italian nationalist movement, the *Risorgimento*, began soon after the Napoleonic wars; for five decades, it was marked by conspiracies, insurrections, rebellions and three separate 'wars of independence', which led to proscriptions sending patriots, and later the partisans of overthrown regimes, on the road to exile. Once again, most sought refuge in Western Europe, London, Paris and Switzerland being the centres of European liberalism (Freitag, 2003). However, we must not forget that, due to its geographical proximity, the most accessible haven for fleeing Italians was the Ottoman Empire. From the Kingdom of Naples and the Papal States on the Adriatic Sea, a quiet boat ride to Albania's coast was easy to arrange. Others travelled via Malta and Corfu to the major cities of the Ottoman Empire where, as discussed above, they found large Italian communities (Michel, 1940). Again, as mentioned earlier with regard to the process of modernizing the Empire, Turkey was keen to attract foreign professionals. Many Italian patriots possessed the necessary qualifications and could be employed as doctors, officers and administrators, or were ready to set up a business. In fact, the opposing groups involved in the *Risorgimento* was reflected in exile to the Orient: on the one hand, liberal partisans of Sardinia and often members of the middle classes and, on the other, the poorer, and sometimes delinquent, Republicans or Mazzinians including deserters for whom it is not clear whether they became Republicans during their exile or that they deserted because they were Republicans.[13]

Finally, we must ask: what role did this complex ensemble of mobilities play during the *Risorgimento* itself, and during the nationalization process in some or all of the Italian communities overseas? Classic historians of the *Risorgimento* have demonstrated that exiles did contribute to the creation of a national awareness: meeting like-minded people in a foreign country was certainly auspicious for discovering shared cultural and political opinions (Maurizio, 2009); however, it was important that they also experienced a sense of loss because of their expulsion. On the other hand, their ability to travel freely between Italy and the Orient and the ease with which they could assume different identities in changing circumstances, as was often the case in the Mediterranean trading community, could also be an indication of indifference to assuming a nationality that was the norm, rather than the exception, until well into the contemporary period, a trend that in France we may tend to forget. Lastly, exile sent patriots, who were educated and highly political in the sense that they had understood the objectives of the

---

[13] ASN, MAE, *busta* 2669 (*Consolati in Grecia*).

national movement, to live in faraway ports where they could act as promoters or sponsors of national identity.

The notion of identity is a recent idea in the public mind: it dates back to the 1960s in the 20th century. It is now omnipresent in the social sciences, but we must maintain our distance from this concept in order to avoid anachronisms. Nevertheless, we can suggest the hypothesis of a gradual transformation of 'being Italian', as the 'Italian question' advanced and regressed over time. Exiles played a role in this process by spreading patriotic ideas. However, most exiles were more interested in surviving and preferred to involve themselves with professional activities rather than political agitation. It is more likely that the Peninsula States tightened the administrative controls on Italian communities in the Orient and that this contributed to the nationalization of their populations. This would explain, on the one hand, the emergence of a more structured consular service and the modernization of European States and, on the other, the development of better organized emigrant communities, notably through the growth of associations which later gave rise to mutual assistance societies. This was a widespread trend during the 19th century but, for Italians in the Orient, it was perhaps more specific to the extent that the various forms of mobility, as discussed above, led to the formation of particularly large communities – for example, several thousand Italians in Istanbul, Izmir and Alexandria – and that the capitulations offered an administrative framework not available elsewhere. The Sardinian, Neapolitan and Tuscan consuls became the leaders of a small, but rather vague, community since it brought together a mixed group of 'nationals': permanent residents registered with the consulate and travellers passing through with a passport, Genoese descendants now known as Sardinians, and local 'dependants' who helped with the community's commercial or administrative activities. Quite apart from the communities, as defined by the administration, there was also a much larger community of 'Italians' that became increasingly compact as the concept of a nation-state grew in popularity in the Orient. Doubtless, co-existence with other Western communities, and in particular the French which was often the best organized and most tightly controlled, contributed to reinforcing a sense of community that went beyond a cultural heritage shared by Italians. After Unification, Sardinian consuls became Italian consuls everywhere and thus responsible for these disparate communities. The community institutions that constituted the consulate, including the newspaper, mutual aid associations, and the hospital, contributed to forging the new nationality during the early phases.

In the decades following Unification, the concept of 'Oriental Italy' became less important since Levantines began to look for support from

France, a country thought to be a more modern and competent protector than Italy where the State was still weak. Administrative procedures for Italian migrants became more standardized and not very dissimilar to those applying to emigrant communities from France, Brazil, the United States or elsewhere. The memory of Italian origins was gradually incorporated into a new administrative programme that sought to maintain links between emigrant 'colonies' and their home country; this programme brought together the State, quasi-governmental agencies such as *Società Dante Alighieri* (Pisa, 1995), and the Catholic Church (Sanfilippo, 2005). In the Eastern Mediterranean, the Ottoman Empire gradually disappeared, whittled away by the combined impact of colonialism and nationalism: cities and regions with Italian populations passed into Western hands, as in Turkey, or were absorbed into local independent countries, notably in the Balkans. The capitulations were no longer relevant and were abolished by the Young Turks in 1908.

References

Andersen Benedict. 1983. *L'imaginaire national. Réflexions sur l'origine et l'essor du nationalisme*. Paris: La Découverte.

Audenino Patrizia and Maddalena Tirabassi. 2008. *Migrazioni italiane: storia e storie dall'ancien régime a oggi*. Milan: Mondadori.

Balard Michel. 1978. *La Romanie génoise. XIIe-début du XVe siècle*. Rome: École française de Rome.

Basso Enrico. 1994. *Genova: un impero sul mare*. Cagliari, Istituto sui rapporti italo-iberici.

Bevilacqua Piero, Andreina De Clementi and Emilio Franzina. 2001. *Storia dell'emigrazione italiana, Vol. 1. Partenze*. Rome: Donzelli Editore.

Bourdieu Pierre. 1980. "L'identité et la représentation. Eléments pour une réflexion critique sur l'idée de région". *Actes de la recherche en sciences sociales*, 35: 63-72.

Carrino Annastella and Biagio Salvemini. 2006. "Porti di campagna, porti di città. Traffici e insediamenti del Regno di Napoli visti da Marsiglia (1710-1846)". *Quaderni storici*, 121-1: 209-254.

Corti Paola and Matteo Sanfilippo (eds.). 2009. *Storia d'Italia. Annali 24. Migrazioni*. Turin: Einaudi.

Dakhlia Jocelyne. 2008. *Lingua franca. Histoire d'une langue métisse en Méditerranée*. Arles: Actes Sud.

Dursteler Eric. 2006. *Venetians in Constantinople. Nation, Identity and Coexistence in the Early Modern Mediterranean*. Baltimore: The Johns Hopkins University Press.

Franzina Emilio. 1994. *Stranieri d'Italia: studi sull' emigrazione italiana dal Risorgimento al fascismo*. Vicenza: Odeon up.

Franzina Emilio. 2006. *Una patria espatriata: lealtà nazionale e caratteri regionali nell'immigrazione italiana all'estero (secoli 19. e 20.)*. Viterbo: Sette citta.

Freitag Sabine (ed.). 2003. *Exiles from European Revolution. Refugees in Mid-Victorian England*. New-York – Oxford: Bergahn Books.

Gabaccia Donna. 2000. *Italy's Many Diasporas*. Seattle: University of Washington Press.

Gellner Ernest. 1999. *Nations et nationalisme*. Paris: Payot.

Genoino Chiara Trara. 1989. "Suonatori ambulanti nelle provincie meridionali: archivio della polizia borbonica e postunitaria dell'ottocento". *La Ricerca Folklorica*, 19: 69-75.

Heyberger Bernard and Chantal Verdeil (eds.). *Hommes de l'entre-deux. Parcours individuels et portraits de groupes sur la frontière de la Méditerranée (XVIe-XXe siècle)*. Paris: Les Indes savantes.

Hobsbawm Eric J. 1990. *Nations et nationalismes depuis 1780*. Paris: Gallimard.

Hobsbawm Eric J. and Terence O. Ranger (eds.). 1983. *L'invention de la tradition*. Paris: Éditions Amsterdam.

Jesné Fabrice. 2009. *Les nationalités balkaniques dans le débat politique italien, de l'Unité au lendemain des guerres balkaniques (1861-1913). Invention scientifique, solidarité méditerranéenne et impérialisme adriatique*. Unpubished PhD Thesis, Université de Paris I.

Lévy Lionel. 1999. *La Nation Juive Portugaise: Livourne, Amsterdam, Tunis, 1591-1951*. Paris: L'Harmattan.

Maurizio Isabella. 2009. *Risorgimento in Exile. Italian Émigrés and the Liberal International in the Post-Napoleonic Era*. Oxford: Oxford University Press.

Michel Ersilio. 1940. "Esuli italiani in Albania (1821-1859)". *Rivista d'Albania*, 1-4: 345-353.

Milano Attilio. 1949. *Storia degli ebrei italiani nel Levante*. Florence: Casa editrice Israel.

Moe Nelson. 1992. "'Altro che Italia!'. Il Sud dei piemontesi (1860-1861)". *Meridiana. Rivista di storia e scienze sociali*, 15: 53-89.

Mola di Nomaglio Gustavo and Roberto Sandri-Giachino. 2007. "La legazione sarda presso la Sublime Porta dal 1815 al 1849". In A. Gasperis and R. Ferrazza (eds.). *Gli italiani di Istanbul. Figure, comunità e istituzioni dalle riforme alla Repubblica 1839-1923*. Turin: Fondazione Giovanni Agnelli.

Nora Pierre. 1984. "Entre mémoire et histoire. La problématique des lieux". In P. Nora (ed.) *Les lieux de mémoire, Vol. 1. La République*. Paris: Gallimard.

Pambianco Filippo. 1926. *Cattolicismo e civiltà italica: agli italiani d'Italia e fuori d'Italia*. Philadelphia: Latin Printing Co.

Petrov Hristo. 2008. "Mobilités du travail (*gurbet*), stratégies sociales et familiales: une étude de cas dans les Balkans centraux". *Balkanologie*, 11-1/2. http://balkanologie.revues.org/index912.html (accessed 12.5.2012).

Pisa Beatrice. 1995. *Nazione e politica nella Società "Dante Alighieri"*. Rome: Bonacci.

Sanfilippo Matteo. 2005. *Dagli indiani agli emigranti. L'attenzione della Chiesa romana al Nuovo Mondo, 1492-1908*. Viterbo: Sette Citta.

Schmitt Oliver Jens. 2007. *Les Levantins. Cadres de vie et identités d'un groupe ethno-confessionnel de l'empire ottoman au 'long' 19e siècle*. Istanbul: Éditions Isis.

Thiesse Anne-Marie. 1999. *La création des identités nationales. Europe XVIIIe-XXe siècle*. Paris: Seuil.

Todorov Tzvetan. 1995. "La mémoire devant l'histoire". *Terrain. Revue d'ethnologie de l'Europe*, 25. http://terrain.revues.org/2854; DOI: 10.4000/terrain.2854 (Accessed 17.5.2012).

Van den Boogert Maurits H. 2005. *The Capitulations and the Ottoman Legal System. Qadis, Consuls and Beraths in the 18th Century*. Leyde – Boston: Brill.

Zucchi John E. 1992. *The Little Slaves of the Harp: Italian Child Street Musicians in Nineteenth-Century Paris, London and New York*. Montreal: McGill-Queen's University Press.

Algerian Jews

How to Make the Return Journey?

Karima DIRÈCHE

If we want to talk of, in today's Algeria, the methods, stakes and impact of a 'return home' by Algerian Jews who left the country after Independence in 1962 or during the 1960s, requires us to question what we should call this population. Categorizing them is a problem because it is linked to different types of belongingness that are used by those involved in returns and to the construction of multi-faceted identities. Since the early 2000s, these 'returns' usually take the form of short-term visits and organized journeys (usually arranged by French or French-Algerian associations, though sometimes by tour operators) and, as such, are linked to the difficult colonial and post-colonial history before and after the War of Independence which left several groups caught in social categories as a result of unspoken assumptions, historical confusions and rejection, on both sides.

The vast majority of Algerian Jews left when Algeria became independent, at the same time as the *Pieds noirs*.[1] Images of the exodus during the summer of 1962 remain ever-present in the public memory: crowded docks in the large port cities or packed departure lounges at Algiers Airport. This context of hurried departures, filled with fear, panic, a feeling of being uprooted, carried with it a series of multiple confusions, misunderstandings, ambiguities and emotions which gave rise, in the years that followed, to a problematic relationship (a combination of nostalgia, rancour and bitterness) with Algeria. It also created memories of an 'Algerian period' marked by war, upheaval and exile. It could be said that

> The hermeneutics of the past for Algerian Jews are no longer purely religious, even though they may be partially constructed on features of religious thought. For most Jews today, their relationship with the past is linked either to the genocide or to a major migration or both (Bahloul, 1992: 199).

---

1 '*Pied noir*' is the term used for a European settler in colonial Algeria. In 1962, there were more than 1 million *pieds noirs* in Algeria, in 15% of whom were Algerian Jews (140,000).

But, above all, Jews have been totally assimilated with the *pieds noirs* and therefore seen as French and, more specifically as repatriated French-Algerian, citizens. Yet, the history of Algerian Jews is not the same as that of *pieds noirs*. Their past in Algeria is socially and historically much older and is shared with the colonized native population, at least until 1870, and their roots reach back to a much earlier era (Ayyoun and Cohen, 1989).

Return journeys to Algeria (whether linked to family nostalgia, pilgrimage, transmission of family history, reconciliation with the past, or appeasement) challenge a number of categorizations and representations of how the Jews established their identity (or what they believed to their identity) as Algerians. To a certain extent, the *pieds noirs'* desire to return was understandable for categories of individuals or groups who could be identified with the period of French colonization since 1830. But the same desire on the part of the Jews is a tangle of different and crossed belongings which makes it much more difficult to identify the different strands.

Can we affirm, today, that Algerian Jews and *pieds noirs* have different expectations when deciding to return or discover Algeria? Especially for those who left when they were very young or born in France and have heard family stories throughout their lives? What is at stake, for them and for Algerians when they return to the country, whether for a pilgrimage, tourism, or a return to their roots? How do they present themselves to Algerians? What impact does returning have on representations of self (whether individual, family or community)?

These issues will be discussed from three perspectives:

1) the different categories: Jews, Israelis, *pieds noirs*, French Algerians, repatriated citizens, for which we will briefly review the history of the Jews in pre-colonial and colonial Algeria;

2) the reasons for the voyage: is this a form of memorial or roots tourism, as seen in Tunisia and, more particularly, Morocco? Or are these journeys to Algeria part of a different approach?

3) Algerian reactions to these returns.

Difficult categories: Jews, *pieds noirs*, Israelites, French of Algeria, the repatriated

The Algerian press, when it publishes articles on visits by Algerian French, usually refers to the return of *pieds noirs* and rarely mentions returns by Algerian Jews; not identifying the religion of people who make pilgrimages to sacred sites such as Tlemcen and Constantine creates a twofold problem:

– For persons returning to Algeria, who tend to call themselves as *pieds noirs* or Algerian French, but very rarely as Algerian Jews;

– For the Algerian government which presents, via the media, these returns as similar to those by Algerian French and, more generally, as an indication of appeasement or as a stabilization of Franco-Algerian relations and as such contributing to historical reconciliation.

These difficulties can undoubtedly be attributed to the status of Algerian Jews which was changed in 1870 with the promulgation of the Crémieux Decree, thus transforming their place in Algerian society (Abitbol, 1983). The Jews were now removed from the category of 'natives' and became French citizens with rights under the French Republic. For a long time, they were referred to as 'Israelites', but this designation did not necessarily acknowledge their roots within the Arab-Berber autochthonous population (which was denigrated and thought to be archaic) since this was now used exclusively for Muslims. For the decades following implementation of the Crémieux decree, Algerian Jews were expected to undergo an intense process of assimilation and francization. It would appear that, on the contrary, in certain social milieux and regions until the 1950s, the majority of Algerian Jews continued to speak in Algerian Arabic or Berber and continued to respect their traditional religious practices (Bahloul, 1992). This has challenged a commonly-held view that the Jews were very rapidly absorbed into the French community. However, two major traumatic events accelerated their francization: the anti-Jewish riots in Constantine on 5 August 1934 (Ageron, 1973) which was almost immediately called a *pogrom*,[2] and the cancellation of their French citizenship by the Vichy

2 "[…] Early in 1930, my father was going shopping. He took me to a neighbourhood very different from that of my school, the lower Kasbah. When we left the market by Rue de la Lyre, Rue Randon (now Rue Amara Ali) led to the Grand Synagogue of Algiers and then became Rue Marengo (now Rue Arbadji). […] what struck me most was the way the people were dressed: older fairly plump women in brightly coloured dresses down to their ankles belted at the waist with a fringed shawl with slippers and another shawl around their heads; most of the men wore in black with a black wide-brimmed hat. I was curious, so I asked my father who these people were. 'They are Jews […], he said, who live in this neighbourhood alongside the native population, with whom they have good relations, but separately from the Europeans. They respect their ancestral customs and wear traditional clothes.' It is true that there was anti-Semitism in the better districts of the city, as I could see from the graffiti on the walls of the University near my high school. Five years later, early in 1935, I returned to this neighbourhood and saw that it had changed completely: the traditional clothes had disappeared, everyone was now wearing European clothes as in other neighbourhoods. These changes must have taken place towards the end of 1934, following a terrible tragedy, a pogrom in Constantine" (Benallègue-Nourredine, 2007).

regime's anti-Jewish legislation.[3] Even today, many of the Algerian Jews who emigrated to France remain silent about this period of exclusion from French society.

A change in attitude certainly occurred at the time of these events and their relations with the Muslim population broke down completely, even though witnesses to these events probably did not really understand that they were being amputated from the indigenous community. 'Francization', even if it was initially denounced by the rabbinic schools because they saw this as a serious setback for their Jewish identity, was soon seen as offering certain advantages and opened up opportunities for social mobility. Taking French first names, adopting French clothes and lifestyles, accepting French values relating to scholastic success: these moves allowed them to differentiate themselves gradually from the Muslim population with whom they had previously shared a common environment. Now Jews and Muslims in the same socio-economic categories moved further apart and cultural and linguistic differences became increasingly marked with their francization. Mentalities changed, Jews had access to schools, jobs and the professions, while enjoying freedoms not available to Muslims: all this drove a wedge between the two communities. In obtaining French citizenship with all the attendant rights, they became part of the privileged classes and were entitled to privileges denied to Muslims who continued to be administered by the *Code de l'indigénat*[4] (Kateb, 2001).

Promotion of Algerian Jews and their integration as French citizens certainly contributed to sentiments of injustice and inequality and further exacerbated tensions between the two populations. The pogrom in Constantine on 5 August 1934 was attributed to the rapid assimilation of Jews in a context of virulent anti-Semitism among French citizens which had encouraged rancour and launched a series of unprecedented and violent episodes. But perhaps the increased Francization, for the Jews, also was linked to their situation as *dhimmis*[5] under Islamic rule? The status of Jews in Islamic countries was neither valued, nor given

---

3 Laws introduced during the period 7.10.1940-20.10.1943, which repealed the Crémieux Decree and created a Commissariat for Jewish Affairs. These were followed by a series of discriminatory measures: revocation of Jewish public servants and a *numerus clausus* in schools and universities. In addition, forced labour was instituted in South Algerian camps.

4 The *Code de l'indigénat* was a collection of discriminatory laws adopted in 1881. It made a distinction between French citizens (from the mainland) and French subjects (from French colonies) and limited the rights and freedoms of colonized peoples. These laws were not repealed until 1946.

5 Term used, under Muslim law, to refer to the 'peoples of the Book'. *Dhimmi* was the status accorded to Jews and Christians and required them to pay additional taxes and respect a certain number of rules and constraints.

value. Arabic terms used for insulting 'the Jew' are very eloquent and reveal the discrimination and often disdain towards Jewish communities in North Africa.

During periods of war, struggle for independence, and Arab-Muslim nationalism, Algerian Jews gradually became *Europeans* to the point where they were seen (often with their acquiescence) as *pieds noirs* (Stora, 2006). Even though they were still called 'israelites', this term no longer had the same connotation as '*indigène*' after the Second World War.

Long considered by Algerians as second-class French citizens (but we must not forget that their rights were withdrawn by Pétain and many Algerian French were anti-Semitic) or a privileged class, Algerian Jews always felt a boundless gratitude to France. Benjamin Stora stated that:

> In regaining their French citizenship after the re-issue of the Crémieux Decree, the vast majority of Algerian Jews knew that they did not want to return to the status of 'native', because, paradoxically, the short-lived Vichy government had reinforced their attachment to French nationality (Stora, 2006: 132).

Misunderstandings, stereotypes, and bitterness were nurtured on both sides. Was the history of Algerian Jews therefore one of those '*Histoires à ne pas dire'* (*Hidden Stories*), the title of a documentary by Jean-Pierre LLedo?[6] After all, what do people really know about the history of Jewish-Muslim relations in Algeria today? According to the economist and author Jacques Attali:

> Algerian Jews were totally cut off from their roots after Independence in 1962. I feel this with sadness, sometimes with bitterness, and sometimes with anger. At a very early date, my father was convinced that Algeria would become independent. But he was also convinced, with sadness, that the Jews would not have a place in an independent Algeria.[7]

This was indeed a painful history, constructed from multiple denials, but this should not allow us, nevertheless, to shirk our responsibility to reflect and ask numerous questions about the reality of the history shared by Jews and Muslims, both before and after the colonial conquest.[8]

---

6 *Histoires à ne pas dire*, Documentary film directed by Jean-Pierre LLedo, released in 2008. LLedo, an Algerian Jew living in exile in France during the 1990s, made contact with four Algerians living in four different regions. They talked about their relationships with Jewish and Christian neighbours during the last years of the colonial period (1955-1962). The film was censored in Algeria.

7 *El Watan*, 3.9.2009.

8 For example, a very long article published on 24.6.2004 in *Le Quotidien d'Oran* (a French-language weekly magazine with a large readership) by a French Algerian

It was not only their indigenous Jewishness during the pre-Crémieux period that had been forgotten. There was also the trauma caused by Pétain's anti-Semite laws (rarely mentioned), which applied to Algeria towards the end of 1940, and the loss of their French citizenship in 1942 (Cantier, 2002).

Recovering their nationality was the last step in the process of francization and their integration and total assimilation into the *pied noir* community – and this led to the total confusion seen in images of the 1962 exodus. Referring to his family memories, Jacques Attali spoke about the question of forgotten roots:

> My father was born in Algeria. His family had century-old roots in the country with a very strong Jewish community. An Algerian Judaism that was extremely cultivated and intense. A Judaism with long-established Berber origins. [...] My father and my mother spoke Arabic. We were not pieds noirs. For me, the history of my presence in Algeria felt like apartheid; we had no contact at all with Arabs.[9]

For a while, thought to represent an intermediary population, somewhere between 'Europeans' and Muslims, in Algeria, the Jews gradually began to fall into the category of *pieds noirs* so that, today for many families, their memory of their Algerian Jewish past has faded away, leaving merely a vague, yet shared, sense of having lost their roots.

The War of Algerian Independence was another key moment when the Algerian Jews' initial identity was put in question. From the beginning, the independence movement was marked by Pan-Arabism and Pan-Islamism, but the place to be attributed to the Jews in the new Algerian nation was rarely discussed. Caught between their new French nationality (and their patriotic gratitude to the French Republic, despite the trauma of the anti-Semite laws) and the forgotten history of their indigenous Jewishness as a result of their assimilation, they were in no position to choose one camp over another.

The War of Independence took place in a context of an intransigent Algerian nationalism and a large majority of Jews preferred to remain prudent and distant. Later, they were reproached for 'sitting on the fence', which was seen by some nationalist militants as a synonym as 'indifference' to the struggle for independence:

> [...] When will the Europeans, and the Israelites in particular, stop watching what is happening without reacting, without calling for an end to injustice? How long will they continue to live with this insecurity? The Israelites, in

---

Jewish academic (Hubert Hannoun) entitled "La déchirure historique des juifs d'Algérie" ("Historical rupture for Algerian Jews"). A vibrant plea for an Algerian Jewish identity.

9 *El Watan*, 3.9.2009.

> any case, do not seem to see it. Not since José Aboulker[10] courageously took up the defence of Algerians against their oppressors during the events of May 1945, has a single Israelite spoken out. They seem to be indifferent to what is going on around them. (Bekkouche, 2002, 37-38)

In 1956, the Algerian Jewish Committee for Social Studies published a declaration in its journal, *Information*, in which it explained its 'wait-and-see' attitude: Algerian Jews had difficulty positioning themselves alongside the independent movement because:

> [...] In these particularly dramatic times when the gap between the various sections of the population in Algeria has become dangerously wide, the Jews who have lived in this country for more than 2000 years, who are deeply grateful for all that they owe to France, and who are attached to this land where they have worked hard to make their business activities prosper, wish to remain faithful to the vocation that makes them equally close to the two other religious communities of Muslims and Christians [...]. With regard to the Muslim community and despite the unjust tribute paid by so many coreligionists, innocent victims who have died in recent months, we have an obligation to pay homage to the respect and cordial relations that have always existed between Jews and Muslims in Algeria and particularly during the Vichy era [...].

The difficulty of this position was also discussed by Frantz Fanon who considered this to be one of the most violent divisions created by colonialism in Algeria and a deliberate attempt to tear apart the children of a same land. In these circumstances, the Crémieux Decree seemed to be a shameful provocation because it gave Jews the illusion of integration with the French people; an integration that was violently rejected by Pétain's anti-Semitist laws. Nevertheless, for Fanon, the progressive elite of the Algerian Jewish community maintained an unquestioning commitment to independence and believed in their participation in the struggle:

> [...] Jewish intellectuals have spontaneously, either through their involvement in democratic and traditionally anti-colonial parties or in groups of liberals, demonstrated their support for the Algerian cause. Today, Jewish lawyers and doctors who shared the destiny of millions of Algerians in camps and prisons bear witness to the multiracial reality of the Algerian nation. In addition, Algeria's Jewish population has official taken position on independence. [...] Attachment to an artificial French citizenship is an illusion at a time when the young and powerful modern Algerian nation is emerging so quickly. Jews have joined the ranks of Algerians fighting for national inde-

---

[10] José Aboulker, a neurosurgeon and son of an Algerian Jewish doctor, was a leader in the Algerian Resistance network during the Second World War. He was known for his denunciations of the massacres at Sétif and Guelma and for his defence of equal rights for Muslim and French populations in 1945.

> pendence [...]. We know also that, in this shared struggle, Muslims and Jews recognize their common brotherhood and their shared deep and absolute attachment for the Algerian homeland (Fanon, 1959: 120).

The situation was much more complex than these words might suggest and the forced migration of 1962 tragically revealed the multiple confusions, misunderstandings, ambiguities and sentiments that made the history of Algerian Jews so problematic. The War of Independence was a key moment which decided the destiny of Jews and led to their departure from Algeria. For Benjamin Stora,

> [...] it was absolutely impossible to believe that French Algeria was finished. The Jews were so absorbed in their 'Frenchness' that they could not imagine themselves living in a Muslim Algeria, because they feared that they would be forced to take their previous status of dhimmis. Right up until the last moment, they hoped that France would remain, that there would be peace, a miracle, and a new more egalitarian Algeria would emerge (Stora, 2006: 130).

Nevertheless, some Jews were indeed ready to participate in the struggle for independence. In the Algerian Communist Party (PCA), there were many young Jews ready to fight for an Independent Algeria. Pierre Ghenassia was one of these: educated at the Lycée Bugeaud in Algiers,[11] he was 17 years old when he joined the National Liberation Army (ALN) in Ténès (a small town 180 km west of Algiers) in May 1956. He came from a Jewish family in Tetouan where his great-grandfather had been a rabbi, while his father (Roger Ghenassia) was a tax officer and his mother (Nedjma Bensaid) owned a jewellery store in Ténès. He was a member of the Union of the Democratic Youth of Algeria (UJDA) and joined the network *La voix du soldat* led by Lucien Hanoun, another Algerian Jew in the PCA and a literature teacher. Pierre Ghenassia died at the age of 18 (22 February 1957) during a bombardment by the French Army in the Blidan Atlas mountains east of Algiers. Until 1990, there was a street named after him in Ténès; but, following pressure from the Islamists, the street was renamed *El Qods* (Jerusalem). Although some former members of the ALN were scandalized by this move, the decision has not been rescinded (Benrabah, 2008).

Established by a small group of intellectuals[12] in December 1956, the Algerian Jewish Committee for the Independence of Algeria also took a strong position in favour of the fight for independence, although they were in a minority amongst Algerian Jews. Its founder, Claude Sixou, was undoubtedly the best known member of the group because he was

---

11 Now the Lycée Emir Abdelkader.

12 Claude Sixou, Pierre Benkemoun, Jean Allouche, Pierre Attal, and André Akem.

the most radical, having deserted and fought in the *maquis*. In 1956, he wrote to the French President, René Coty,[13] explaining why his desire to be an Algerian which he believed the colonial system had confiscated:

> The deferral accorded to me by the military authorities so that I can complete my studies has just expired. Any day now, I will be summoned to carry out my military service. I must inform you that I will not comply with any summons and explain to you in detail the reasons for my attitude: I was born in Tiaret in Algeria on 11 June 1931 and I have lived more than twenty years of my life here. At an early age, I had an opportunity of seeing for myself the falseness of statements made by French governments that Algeria would be a French province. It was when I came to France for the first time and I talked with French democrats that I began to realize that there were profound reasons for the evils afflicting my country. I became convinced, and have seen no evidence to dissuade me from this position, that it was possible to establish a free and happy Algeria, on condition that the odious colonial regime that was imposed on that country more than 120 years ago is defeated.

His total commitment to independence naturally led him to apply for Algerian naturalization in 1962, but it was refused under the new Code for Nationality which did not recognize (and still does not recognize today) Algerian citizenship for non-Muslim militants. This denial of recognition was the cause of real suffering not only for Sixou, but also for the small group of Jewish partisans who had chosen Algeria as their homeland.[14]

When they left for France in the terrible conditions described above, Algerian Jews were treated in exactly the same way as the Algerian French and received the same indemnities as all other groups repatriated from Algeria.[15] It is thought that there are 400,000 Algerian Jews in the Jewish community in France today. Unlike Moroccan and Tunisian Jews, many of whom emigrated in Israel in large numbers, the vast majority of Algerian Jews settled in France, even though the efforts made by the Universal Israelite Alliance and the establishment of the Jewish Foyer in Palestine had already encouraged them to migrate after the creation of the State of Israel in 1948.

---

13 Letter reprinted on 15 June 2011 in http://www.lematindz.net/news.

14 *El Watan* published an announcement of Claude Sixou's death: "Mohammed Harbi and the journal NAQD have informed us with great sorrow of the death, on Friday 10 June 2011, of *moudjahid* Claude Sixou, founder of the Committee of Algerian Jews for the Independence of Algeria in December 1956 and director of the journal *Sou'al* in the 1980s. On this sad occasion, they wish to present their most sincere condolences to the family of our comrade in arms".

15 Four separate laws for compensation: Law No. 70-632 (15.7.1970); Law No. 78-1 (2.1.1978); Law 82-4 (6.1.1982); Law No. 87-549 (16.7.1987).

Thus, we can see that the history of Algerian Jews is not the same as that of the *pieds noirs*. It is part of a different historical perspective and of a history that is shared with the Algerians but covers more than the colonial period (Abécassis and Dirèche, 2012: 141-151).

Why do Algerian Jews return?

Compared to the situation in Morocco and Tunisia (Fellous, 2003), there is no specifically Jewish tourism to Algeria. For example, although Moroccan Sephardis emigrated to Israel en masse, this has not in any way effaced a deep attachment for their country of origin. On the contrary, the number of religious pilgrimages organized by the Moroccan Jewish diaspora is growing each year. Nearly 5,000 Jews (mostly Israelis) travel each year on a pilgrimage to one of the 1,200 Jewish sanctuaries in Morocco. This religious tourism is encouraged and promoted by Morocco which, for a number of years, has maintained the heritage of Jewish identity in the country (Dakhama, 2005). Cultural events and academic conferences[16] on this theme allow us to see that Morocco has accepted the religious diversity of its past, even if the political and social realities are somewhat problematic. Furthermore, the highest institutions in the country have taken an official position on the history of Judeo-Islamic cohabitation that is specific to Morocco. Such statements cannot be made of Algeria where references to the Jewish community are avoided or even denied, sublimated by a strict and ideological reinterpretation of this part of Algeria's history: that of a religious group that is now totally extinct, but which benefited from 'colonial advantages' due to the Crémieux Decree and has 'forgotten' its indigenous history. The Israel-Palestine conflict and the political attitude of the Algerian government have rendered this position more radical so that, today, any discussion of the Algerian Jewish community is very often met with suspicion and intransigence.

Even if it has always been possible to make personal visits to Algeria, group travel is a more recent phenomenon, especially during the 2000s after the election of Abdelaziz Bouteflika as president (1999). Bouteflika was one of the principal actors in improving the political situation in the country by ending the unacknowledged civil war (known as the *black decade*) that caused so much grief to Algerians during the 1990s. The promulgation of laws for amnesty and national reconciliation in 2005 encouraged hopes that the country's political future would now be more serene.

---

16 For example, a seminar on "Le judaïsme contemporain et le Maroc de demain", held in Casablanca in 2008, and a special section in the magazine *Zamane* (*Time* in Arabic) on "Why and how did Morocco lose its Jews?" in its first issue.

The first return journeys by the Jewish community took place much earlier as part of week-long celebrations of Jewish-Algerian friendship; a first voyage was organized in 1978 with visits to Tlemcen and Nedroma in the western region of Algeria. Both towns have a number of historical Jewish sanctuaries and are major pilgrimage destinations. At Tlemcen, there is the famous tomb of Rabbi Ephraim N'Kaoua, renowned for his miracles throughout North Africa,[17] while Nédroma is home to the sanctuary of Sidna Yousha (or Yoshué) which attracts pilgrims from Morocco and Tunisia. These are high places in Jewish sainthood in North Africa. A documentary film[18] demonstrates the importance of this return journey for the community 16 years after Independence. The camera focuses on the extreme intensity of the pilgrims' rediscovery[19] of their birthplace and the visit to the cemetery is particularly moving as pilgrims search for the tombs of their dead with emotion, tears and prayers.[20]

It was not until May 2005 that a second voyage by the Jewish community (130 persons) was organized with assistance from the Algerian Embassy in France and the association La Fraternelle (which has 1,300 Jewish members originally from Tlemcen). This was an official voyage which made the front pages of the Algerian press and which, through the presence of these pilgrims, encouraged hopes for improving relations between France and Algeria. This desire for a stable political relationship with France allowed the Algerian authorities to acknowledge the many difficulties in recognizing Algerian Jews: were they children of the country? Or *pieds noirs*? It was clear that any reference to their status as autochthones was not on the agenda and that it was better to speak of them as repatriated Algerian French.

Tlemcen (where there is a neighbourhood still called *derb lihoud*[21] with 3 abandoned synagogues) is the only Algerian town to have offered an official welcome to members of an organized Jewish pilgrimage. However, the Gaza War in January 2009 brought these journeys to a sudden end and pilgrimages were no longer tolerated.

---

17 According to legend, he fled after the Spanish *Reconquista* and sought asylum from the Sultan of Tlemcen.

18 www.juif.org/video/3080/perinage-des-juifs-de-tlemcen.

19 Identifying places in the country ("this is my neighbourhood, my house; it's the third on the left, the one with a balcony"), finding traces of their existence, their despair ("I can't see anything I recognize"; "there's nothing left; all the buildings and houses have been destroyed"); recognition of neighbours and friends.

20 The film is very emotional and the scenes showing the washing of tombs are particularly moving.

21 'Jewish quarter' in Arabic.

Thus, the presence of groups of Jewish travellers seemed to embarrass Algerian politicians, since the presence (and recognition) of Jews was closely linked to events in international politics and the conflict between Israel and Palestine.

Travel agencies specializing in 'return packages' and associations are now organizing these journeys to Algeria. However, these tend to minimize the Jewish theme by concentrating on a more general approach, either cultural or associative.

APF France-Maghreb, a French association, has often taken responsibility for organizing journeys by *pieds noirs*. Initially, its role was to renovate and rehabilitate the French cemeteries to be visited during these journeys, but it later extended its activities to various regions in the country. APF France-Maghreb, in partnership with Algerian associations and a children's heart surgery service, then launched humanitarian activities relating to campaigns to reduce perinatal deaths. Among the Algerian French participants, there were many Jews who saw this as an opportunity to make a return trip which would allow them to take advantage of the package tour while downplaying the religious aspects which they did not necessarily wish to make public.

In 2005 and again in 2007, a debate in the Algerian press over demands by *pieds noirs* for compensation for lost property and assets, valued at €14 million, from Algeria,[22] highlighted the confusion that today's Algerians have as to the identity and 'historic' grievances of Algerian Jews. This issue was taken up by *Le Quotidien d'Oran*, a French-language magazine,[23] and focused particularly on demands for compensation supposedly made by Jewish families. It pointed out that

> they [Algerian Jews] had accepted the Crémieux Decree during the colonial period and had decided to leave for France at the same time as the pieds noirs and the harkis[24] on the pretext that they were afraid of reprisals.

22 Amount demanded by the Union syndicale de défense des Intérêts des Français repliés d'Algérie et d'Outre-Mer (USDIFRA) and the Groupement national pour l'indemnisation des biens spoliés ou perdus outre-mer (GNPI). These associations claim the French Republic, despite clauses in the Evian Accords, had not honoured its fundamental obligations, as they felt that the various laws on compensation laws assessed the value of 'despoiled' property at only 30-40% of its true value. On the other hand, the Fondation du 8 mai 45 claimed compensation of $100 billion from France for prejudices suffered during 130 years of colonization.

23 *Le Quotidien d'Oran*, 12 May 2005, "Les juifs d'Algérie demandent 144 millions de dollars".

24 The *harkis* were Algerian auxiliary soldiers serving in the French army between 1957 and 1962. Seen as traitors by the Algerians, many of them were able to leave the country when it became independent and found refuge in France.

The Arabic newspaper, *Al Quds*, which is published in London, echoed this sentiment but did not cite its sources. The strident nature of this debate in the national press was influenced by other international events: Israel had drafted legislation known as the 'Law for the Compensation of Jewish Refugees from Arab Countries'. This legislation was adopted by the Knesset in February 2010 and gave Jews who had fled from persecution in certain Arab countries the status of 'refugees'. For Israel, this was a way of restoring 'historical justice'. Algeria was the first country to react, stating that it would not recognize any claims for compensation from Israel.

This debate in the national press is interesting for several reasons, as it highlights dysfunctional understandings, difficulties in identifying actors and confusion of issues. Here Jews were not identified as *pieds noirs* but as Jews backed by Israel. *Le Quotidien d'Oran* noted that:

> unlike the pieds noirs who have expressed their desire for return visits to the places of their childhood, Algerian Jews have never asked for anything of the sort.[25]

How are Jews represented and how are they received?

References to the hospitality, warmth, kindness and welcome from the Algerian population, as well as the evocation of rediscoveries and memories of the dear departed (very emotionally upsetting for the visitors), offer a more nuanced approach to the political issues involved in the representations we have just described. Algerians had kept alive their warm memories of their Jewish neighbours and the discovery of the Arabs' respect for their dead stirred up emotional reactions from the visitors themselves. This was a lesson in fraternity, courtesy and humility,[26] offered by the Algerians and mentioned by many travellers, which was markedly different from the official view of Algeria's history as presented in the media and school textbooks. And even if the past is (always?) reconstructed *a posteriori*,[27] as an idealized representation of relationships between Jewish and Muslim communities, the intense emotions awakened in rediscovering neighbours were nevertheless a strong indication of a shared and common memory. The impact of the departure of Algerian Jews, on both communities, can be traced back to

25 *Idem*.

26 See articles available on the following sites http://piednoirs.viabloga.com (visited in September 2009) and http://zlabia.com.

27 For one pilgrim, the warm welcome from inhabitants of Nedroma was surprising: "Before, this didn't happen. We would rarely see a Muslim kiss or embrace an Israelite. We would nod our head to say hello. You would never see an Israelite kiss a Muslim". www.juif.org/video/3080/pelerinage-des-juifs-de-tlemcen.

the 1870 Crémieux Decree (which removed Jews from the category of indigenous inhabitants and forced them to take on different identities) has not been sufficiently studied in terms of the violent breakdown of inter-community relationships which led to amputation from the wider society. This was a major trauma for Algerian identity, admirably expressed by the philosopher Jacques Derrida in a text entitled *Moi, l'Algérien* in 2007:[28]

> I want to speak as an Algerian, born an Algerian Jew, about this section of the community which, thanks to the Crémieux Decree in 1870, received French nationality but lost it again in 1940. When I was 10 years old, I lost my French citizenship under the Vichy regime and, for several years, was forbidden to attend French schools. I was one of the people they called Algerian Jews at the time. It was one of the great events of my whole existence. There were others. [...] I speak here as an Algerian who became French at a given moment in time, lost his French citizenship and then found it again. [...]. A Jewish/French/Maghreb genealogy that cannot explain everything, far from it, but how can I explain anything in my life without it?

How is this seen from the Algerian side? Little has been written on the subject and all too often such opinions are buried under other discordant and often contradictory arguments. In 2008, *La Tribune* published an article entitled "Jewish questions *à l'algérienne*" (Bouhamidi, 2008) which was one of the few attempts to reflect on the upheavals experienced by Jews in Algeria after promulgation of the Crémieux Decree: a real 'fracture line' which finally forced them into colonized society, but on the side of the dominant class. It was a profound change but also offered a chance of upward social mobility, though at the cost of an almost definitive rupture with their Muslim neighbours. The author also mentions Algerian anti-Semitism in the following terms:

> [...] in their contacts with Jews, Algerian Muslims continued to refer to old medieval controversies on the superiority of certain religions over others. As subjects of multi-ethnic, multi-cultural and multi-linguistic Muslim empires, Jews, Christians and Muslims were continually at odds over the true faith. Nothing authorized their involvement tin religious contests on the basis of race since Christians and Muslims are often of the same race in today's Palestine or Lebanon. Religious hostility focused mainly on religious characteristics and this explains the absence of pogroms in Arab countries.

---

28 http://www.lematindz.net/news/373-lanti-macias-moi-lalgerien-de-jacques-derrida.html. Jacques Derrida's comments were made during an acrimonious debate over a proposal that Enrico Macias should return to Algeria and give concerts. Macias is a Jewish singer from Constantine whose songs tell of exile and rootlessness are very popular with the *pieds noirs*. He was twice refused entry into Algeria (in 1999 and 2007) because of his pro-Zionist opinions.

This is an interesting and courageous attempt to reflect on the situation but it still sidesteps the ages-old debate on social and inter-religious relations between Muslims and Jews and continues to spread the false idea that pogroms never occurred in Muslim countries.[29] On the other hand, the text enters into a long discussion on anti-Semitism as practised by Algerian Europeans which reached its apogee with Pétain's anti-Semitic laws which cancelled their French citizenship. As a result, Algerian Jews were now seen as victims at two levels: second-class French citizens and assimilated persons stripped of their initial identity, while, regardless of the humiliations and rejection they had suffered at the hands of the Vichy government, they wanted desperately to be French. This aspect of the history of Algerian Judaism is still a taboo subject. In the current international context and attitudes towards the Israel-Palestine conflict, it has been ignored, in part because of confused notions about anti-Zionism and anti-Semitism throughout the Muslim world.

The singular destiny of Algerian Jews … One day, no doubt, it may be possible to study the impact of the Crémieux Decree and the exodus of 1962 as events that led to an amputation of Algerian society: an amputation accompanied by sufferings, scars and opportunities for revising history.

It is a history that has not been told, that is not part of expressions of self and that only partly explains the confusions and ambiguities of the telling and interpretation of this very history. For millennia, Algerian Jews have lived astride two worlds, yet are caught in an in-between environment of great complexity due to the intricate links that bind Jew, Berber and Arab in a single population yet, for too long, conscious of their dual culture: they are without a doubt *frères de terre.*[30] Yet, the winds of change that influence world history have led to the gradual loss of their mother tongue and of a shared linguistic and cultural symbols, while distance from their ancestral lands has made them strangers in the land of their birth.

References

Abécassis F. and K. Dirèche (eds.). 2012. *La bienvenue et l'adieu. Migrants juifs et musulmans au Maghreb, XVe et XXe siècles*. Casablanca: Éditions La croisée des chemins/Karthala.

Abitbol M. 1983. *Les juifs d'Afrique du Nord sous Vichy*. Paris: Maisonneuve & Larose.

[29] There were several other pogroms during the 20th century: Fès in 1912, Palestine between 1920 and 1929, Constantine in 1934, Bagdad in 1941, etc.

[30] Brothers through their connection with the land – translator's note.

Ageron C.-R. 1973. " Une émeute anti-juive à Constantine (août 1934)". *Revue de l'Occident Musulman et de la Méditerranée*, 13-14: 23-40.

Allouche-Benayoun J. and D. Bensimon. 1989. *Les juifs d'Algérie. Mémoires et identités plurielles*. Paris: Éditions Statvit/Cerf.

Attal R. 1996. *Regards sur les juifs d'Algérie*. Paris: L'Harmattan.

Ayyoun R. and B. Cohen. 1989. *Les juifs d'Algérie, deux mille ans d'histoire*. Paris: Jean-Claude Lattès.

Bahloul J. 1992. *La maison de mémoire. Ethnologie d'une demeure Judéo-arabe en Algérie, 1937-1961*. Paris: Métailié.

Bekkouche M. 2002. *Journal d'un oublié*. Alger: ANEP.

Benallègue-Nourredine A. 2007. *Le devoir d'espérance*. Algiers: Casbah Éditions.

Benrabah M. "Mort au champ d'honneur le 22 février 1957". *Alger républicain*, May-June issue.

Bouhamidi M. 2008. "Questions juives à l'algérienne". *La Tribune*, 24.4.2008, also available on the website www.algeria-watch.org.

Cantier J. 2002. *L'Algérie sous le régime de Vichy*. Paris: Odile Jacob.

Dakhama H. 2005. "La présentation de soi au pèlerinage de Rabbi Hahya Lakhdar (Maroc)", in S. Chiffoleau and A. Madoeuf (eds.), *Les pélerinages au Maghreb et au Moyen-Orient. Espaces publics, espaces du public*, Beirut, Ifpo: 39-58.

Dirèche K. 2008. *Prosper-Messaoud. Un juif d'Oran raconte*. DVD. Rabat: Éditions du CJB.

Eisenbeth M. 1936. *Les juifs de l'Afrique du nord. Démographie et onomastique*. Algiers, Imprimerie du Lycée.

Fanon F. 1959. *L'an V de la révolution algérienne*. Pari: Maspero.

Fellous S. (ed.). 2003. *Juifs et musulmans de Tunisie. Fraternité et déchirements*, Paris: Somagy.

Kateb K. 2001. *Européens, indigènes et Juifs en Algérie, (1830-1962)*, Paris: Travaux et Documents, INED/PUF.

Stora B. 2006. *Les trois exils. Juifs d'Algérie*. Paris: Stock.

"Invisibility" and Memory

Italian Immigration in France During the Second Half of the 20th Century

Stéphane MOURLANE, Céline REGNARD

There is a long history of Italian immigration in France. Since the Middle Ages, Italians have continually crossed the Alps, often in search of seasonal agricultural work or commercial opportunities in the South (Corti and Sanfillipo, 2009). With a constant arrival of migrants, frontier regions (Alpine slopes and the South-East) became traditional settlement areas for Italians migrants. After the 19th century's Industrial Revolution, the number of immigrants rose and became more substantial for two major reasons: substantial differences in population growth and in business development opportunities between the two countries. During this period, waves of migrants settled not only in major cities but also in the eastern part of the country, thus reinforcing the importance of these frontier regions.

Italian migrants settled mainly in the South-East, Paris and the North East, and in particular Lorraine where, by 2004, they had become the largest group of foreigners (Galloro, 2007). Two factors – the cities and closeness to the frontier – led to their settlement in regions where industrial jobs were available from the late 19th century onwards.

The arrival of migrants continued for more than a century until the early 1960s, with three distinct phases (Bevilacqua *et al.*, 2002).

The first phase was known in Italy as "the great emigration", beginning in the late 19th century and lasting until the end of the First World War. Just before the War, there were 420,000 Italians in France, approximately 36% of all foreign residents. At this point, France was the third destination for Italian migrants, after the United States and Argentina.

The second phase took place between the two world wars. In France, the high loss of men during the war created a lack of labourers while the economic crisis in Italy, together with the ascension of Mussolini, led to more people seeking to emigrate. Another important factor was the introduction of new US legislation which imposed quotas (in 1921 and 1924) and therefore forced many migrants to seek other destinations. In 1924, 200,000 Italians left Italy for France. By 1931, there were

1 million Italians (out of 3 million foreigners), representing 7% of the total population.

The number of Italian migrants dropped during the Second World War: there were 450,000 Italians living in France in 1946 and this marked the beginning of the third and final phase. Between 1945 and 1949, 67% of legal migrants were Italians (Tapinos, 1975). In March 1951, France and Italy signed an agreement which would regulate and provide structure for future arrivals of migrants (Mourlane, 2002). The number of Italians in France had grown to 570,000 by 1968. At this stage, Italians were still the largest foreign population in France, but the situation was about to change. After 1962, the flow of migrants began to fall: Italy was experiencing an "economic miracle" and this new-found growth meant that fewer people wanted to leave and emigrants were more interested (especially Southern Italians) in countries with better salaries (Blanc-Chaléard, 2003).

By the end of the 20th century, it was estimated that there were between 3.5 and 4 million people of Italian origin living in France, out of a total population of 60 million (Milza, 1993). According to Tribalat (1991), some 10 to 11 million foreigners (migrants and their descendants) were living in France since the end of the 19th century: Italians therefore represented a large proportion of this group.

And yet, over the last 50 years, whenever the media discussed the question of migration, the debate focused on migrants who arrived more recently, in particular North Africans. Italian immigration seemed to disappear from public memory and become invisible; however, it would seem that, in recent years, there has been a revival, albeit in different forms.

This article will discuss and analyze the three different ways in which memory of Italian immigration in France, and in particular in frontier regions, has evolved: amnesia, revival, and commemoration.

Invisibility and amnesia (1950-1960)

The first waves of Italian migrants at the end of the 19th century were not given a particularly enthusiastic welcome. Due to tensions between France and Italy and difficulties in the French labour market, where Italians were seen as unfair competition, they were treated with resentment which often led to violence (Milza, 1981; Dornel, 2004). A climate of xenophobia, nurtured on myths of invasion, also created an environment that encouraged the emergence of unflattering stereotypes (Schor, 1985). Although the South-East has a long history of migration and relations between the French and Italians have been close for centuries

(Faidutti-Rudoph, 1964; Corti, 2003), this frontier region has not always had a good reputation for welcoming migrants, far from it.

The development of a well-defined national space coincided with the first waves of Italian migrants at the end of the 19th century and this led to their becoming an issue in the debate on 'nation' in these regions (Noiriel, 2001), as in others, because of their impact on the economy, as they created competition in the labour force, and on the political situation, by challenging national identity, and particularly in the County of Nice after it became part of France in 1860 (Schor *et al.*, 2010). It is, after all, always easier to have a debate on the concept of nation when the question of difference can be linked to close cultural proximity. For this reason Italians were often rejected in these frontier regions, as were Belgians in the North (Dornel, 2004; Sanna, 2011).

After the Second World War, France remained very negative towards Italians, particularly in the South-East (Gastaut, 1996), where the presence of an Italian occupation army during the war had contributed to reviving anti-Italian feelings (Panicacci, 2010). Within a few years, however, French opinion appeared to have forgotten the existence of Italians.

In his PhD thesis on public opinion and immigration during the 5th Republic, Yvan Gastaut (2000) referred of the "invisibility of European migrants since the 1960s". Even if some prejudices still remained (Mourlane, 2007), attitudes towards Italians nevertheless seemed to be more favourable. In an INED study carried out in 1951, Italians were ranked fourth, in terms of positive attitudes towards migrants, after Belgians, Swiss and Dutch, and higher than the Spanish and Polish contingents (Girard and Stoetzel, 1954). According to a 1966 opinion poll, 61% of respondents had a good opinion of Italians while 24% expressed a poor opinion. Another survey, carried out in the same year, asked which nationalities were thought to be the most numerous: Italians were not mentioned among the responses. It would appear that the long history of Italian migrants in the country worked to their advantage, despite the fact that, in the 1968 census, half of the Italian residents who gave their date of entry had arrived in France since 1956, and three-quarters since 1946. This suggests a discrepancy between their relatively recent arrival and perceptions of their long-established presence in France which contributed to their 'invisibility'. At the same time, manifestations of hostility had not completely disappeared. During the Algerian War, there were many reports of a tendency to assimilate Italians with Algerians (Blanc-Chaléard, 2000). It is common for Italian children who immigrated to France during the 1950s to mention instances of taunting in school playgrounds. Despite a few loose expressions stigmatizing their origins, we can say that, by the 1960s, Italians were no longer victims of xenophobia. We can thus see a gap between the reality and the perception of their presence in France. This distortion in public opinion can be attributed mainly to traditional feelings of cultural affinity. Italians were seen, according to the studies carried out by Jean-Charles Vegliante, as "almost the same [as us]" (Vegliante, 1995). The similarities observed are therefore part of a common inherited culture, as can be seen in references to Italy as our "Latin sister".

The situation changed during the 1960s: the end of Italian immigration, in large part thanks to the Italian economic miracle, coincided with the beginning of a wave of Algerian migration, which increased substantially after the Evian accords. References to 'us and them', a common expression of xenophobic attitudes, now focused on Algerians, rather than on Italians (Taguieff, 1988; Gastaut and Mourlane, 2009). In Lorraine, Piero Galloro also observed this phenomenon and mentioned in particular certain social aspects including a "*re-categorization* [that] has occurred in favour of Italians in which the Italian worker's image is assimilated with that of the French worker, sharing many values to the point of resembling them and thus of being accepted as part a new group alongside local workers" (2007, 111).

It should be noted that Italian migrants seem to be somewhat discreet in their dealings with the French. Despite a growing interest in their national identity, as can be seen from higher membership of cultural associations, Italians do not appear to be particularly interested in ostentatious displays of their Italianness. They no longer speak their native language outside the family home. Emile Temime's studies in Marseilles have shown that, in the South-East, Italians adapted more easily to their new language since French remained, until the mid-20th century, very similar to their local dialects (Temime, 1985; Pasquini, 1993). Dominique Schnapper (1974), in an article comparing the situation in France and the USA, noted that, by the late 1960s, Italian migrants could not be distinguished from the French population in public places. Furthermore, they had abandoned their traditional religious practices and dress codes. The last remaining links with their culture were Italian cuisine, enjoyed only within the family circle, and their accent, the only way their origins were still obvious outside the home. As a result of this obsession for transparency, the organization of specifically Italian cultural activities was almost non-existent and their Italianness was gradually absorbed into the dominant socio-cultural environment (Vegliante, 1995).

French media took little interest in Italian immigration (Mourlane, 2002). A quantitative analysis of articles published in *Le Monde*, considered to be a "good way to understand French society" (Thibau, 1996), confirms this impression of understatement, or even absence, of immigration issues. Despite its preoccupation with the impact of the Algerian War on immigration during the period 1958-1962, *Le Monde* was only marginally interested in the difficulties faced by immigrants after 1963. Nevertheless, the proportion of Italians in the population was slowly declining and was almost non-existent after 1967. In the late 1960s, *Le Monde* and the rest of the print media focused their attention on Algerian migrants, with only a cursory interest in European, and then mainly Spanish and Portuguese, migrants.

One of the last films to focus on Italian workers was Yves Allegret's *La meilleure part*, released in 1955. The film portrayed Italians as a hard-working population whose integration went hand-in-hand with their disdainful treatment of later arrivals, Algerian migrants. In the 1960s only one film, Tony Richardson's *Mademoiselle* (a Franco-British coproduction) showed Italians working in an unusual sector: as lumberjacks in the rural department of Corrèze. The originality of this film lies in its evocation of an emerging racism against Italians.

Thus, in the 1960s, two phenomena coalesced and brought an end to any possibility of a collective Italian memory: absence in the media together with a benevolent attitude towards their presence. Both can be explained, and even justified, by the absence of migrant memory.

Rediscovering memory (1970-1990)

It was not until the 1970s that new instances of cultural expressions of Italianness surfaced. An early example was the song *L'Italien* written and recorded in 1971 by a popular and provocative singer, Serge Reggiani, who had arrived in France with his family of opponents to the Fascist regime in 1930 and become a strong supporter of left-wing politics, particularly during the 1968 uprising (Brierre, 2005). *L'Italien*, written and sung in French and Italian, spoke of the migrant who returns to his home country after a difficult journey and of all the difficulties that he has to face and his attachment to his country of origin. It was the main track on an album that sold more than 2.3 million copies and Reggiani regularly included it his concerts right up until the 1980s. Another album, *En Italien*[1] released in Italian in 1972, contributed to Reggiani's role as a standard-bearer for Italians and Italian descendants living in France.

Another Italian migrant to achieve fame in France was François Cavanna who, a few years later in 1978, published *Les Ritals*. He was a cartoonist and one of the founders of two very popular satirical maga-

1 For more details on Serge Reggiani's repertoire, see http://www.serge-reggiani.com (accessed 13.7.2012).

zines, *Hara-Kiri* and *Charlie Hebdo*. In *Les Ritals* (the nickname given to Italians in the 1950s), Cavanna wrote about his childhood during the 1930s in Nogent-sur-Marne where there was a large Italian community (Milza and Blanc-Chaléard, 1995). Well received by the critics, the book revived images of a past history that was almost forgotten by the local community: by now, the Italians had moved on and streets that had been filled with the shacks of Italian migrants were destroyed to make way for new urban planning.

Sport offered a new candidate to epitomize the memory of Italian immigration: Michel Platini, the best football player in France during the 1970s and 1980s (Mourlane, 2003). Since its first match in 1912, the French national football team had always included migrants of various generations, long before its victory in the 1998 World Cup gave rise to "a French passion for multiculturalism" (Gastaut, 2008). Players of Italian origin were often picked for the team: another was Roger Piantoni, famous for his playing with the very successful team of Reims and the national team during the 1958 World Cup in Sweden (Mourlane, 2010).

Like Piantoni, Platini was a third generation Italian whose family had settled in Lorraine, Eastern France, where many Italians were recruited to work in the steel mills. Although he never referred to his origins in public, press coverage of his career often focused on his youth and family history, typical of the life of many Italians in Lorraine: holidays in Italy, conversations in Italian, playing *scopa* (an Italian card game), passion for Italian football, etc.

When he was transferred to the Juventus football club (Turin) in 1982, references of his origins increased. At the time, the news magazine *Paris-Match* wrote: "*his return to his parents' home country became a sentimental journey*" (14 May 1982). However, it seems that Platini experienced a period of identity confusion. Shortly before the 1982 World Cup, in response to a question from a journalist about whether he thought of himself as Italian, he said "I don't know. When I heard the anthems before the France-Italy match, I felt very moved and I asked myself 'who am I?' I was moved by both hymns. I am French, obviously, but in my heart, there is also my blood, my family, my father …" The way that Michel Platini expressed his relationship with his Italian origins can be interpreted as a 'return to source', often observed amongst third generation migrants (Sayad, 1977). But, in this instance, there was a small nuance in that his return was not voluntary, as it was dictated by his sporting career. At the same time, this growing awareness of his Italian connections also strengthened his attachment to France. "My roots are also in France" he told *Paris-Match*.

Public figures like Reggiani, Cavanna, and Platini – and indeed Yves Montand, the actor and singer who was born in Italy but grew up in Marseilles – came to epitomize Italian migrants who, although fully integrated and even assimilated, did not forget their cultural origins and their life as migrants. They came to represent the memory of Italian immigration by the very fact of their public persona. Here again, we can see how their attachment to their home region influenced Italian immigrants and remained strong with the re-emergence of Italianness in French popular culture. Cavanna wrote about his childhood in his home town of Nogent, where there were many Italian immigrants. On the other hand, Platini's insistence on his Italian origins went hand-in-hand with references to his family in Lorraine. He became the symbol of double identity (immigrant and regional) as an Italian-Lorrain, a sentiment not often found in other French regions.

History also has a part to play in the process of reviving memory. As we know, history and memory are very closely and intricately linked, though often in opposition to each other, as suggested by Pierre Nora in his introduction to *Les Lieux de mémoires*: the rationality of universal history is often offset by emotional ties to the individual or collective memory (Nora, 1984-93). And yet it is probably fairer to focus on the dialectic relationship between history and memory. Memory nurtures history, while history contributes to the creation of collective memory.

The history of migration in France developed in the 1980s. At the time, the economic crisis and the end of mass immigration created an increasingly stable situation with the last waves of Italian immigration in France. In 1981, foreigners were accorded the right to join associations and unions and, as a result, a large number of migration-linked associations were founded and cultural activities were encouraged. These associations also benefitted from the public fascination for memory and heritage and this led, in the early days, to the establishment of links between history and immigration.

The other major event was the return of the far right on the French political scene in 1983 (Noiriel, 2005). Not only did the *Front National* make immigration the central focus of its propaganda, it also revived past arguments about national identity, and thus strongly contributed to placing memory at the heart of the immigration debate. One of the far right's main arguments was that immigrants cannot be integrated because they do not share the same culture and history as the 'native' French. This argument sought to make immigration a new problem for French society while, at the same time, emphasizing the idea that the individual's 'roots' are an essential component of their feelings of belonging and of loyalty to the nation. Historians, who had not previously taken a great interest in the question of immigration, suddenly took notice, not least because many had experience of xenophobic behaviour, either personally or in their family (Rygiel, 2009).

The party's electoral results were particularly strong in regions where there was a substantial immigrant or foreign population. Its electoral base in the country corresponded to areas with migration flows and major centres of manufacturing industry. In the South-Eastern departments of Bouches-du-Rhône (Marseille) and Alpes-Maritimes (Nice) in particular, the Front National obtained its earliest electoral victories (Vaughan, 1991).

Gérard Noiriel and Pierre Milza were among the first historians in this field. In his PhD thesis on the relations between *Français et Italiens à la fin du XIXe siècle* ("French and Italians at the end of the 19th century"), Milza focused on Italian immigration in France. As the first Chairman of the Centre d'Études et de Documentation de l'Émigration Italienne, he published many articles and collections of seminar papers. In 1993, he published a book in which memory and history intertwine: *Voyage en Ritalie*. The first chapter, in which he referred to his own search for personal roots as the son of an Italian father and a French mother, opened with the sentence "*I am a migrant, uncomfortably caught between two similar, and utterly dissimilar, cultures ...*" Milza said that, for many years, he knew nothing about Italian culture. It was only later, first as an adolescent and later as a historian, that he decided to visit Italy. His story took on the appearance of an initiation journey as part of his personal history. His decision to focus his research career on Italian immigration was therefore as much dictated by scientific curiosity, as by an attempt to define his own identity. *Voyage en Ritalie* offers a synthesis of what Ralph Schor calls a "scientific autobiography",[2] in which historians immerse themselves in their own memories in order to reflect and draw conclusions. At the same time, the book offered a historical approach which gave a larger non-academic readership access to the memory of Italian immigration and recognized its importance in constructing that memory.

Regardless of the cultural form used in constructing memory, such attempts constitute the expression of two contradictory, but regularly linked, aspects that define Italian immigration: integration and rejection (Colucci and Sanfilippo, 2010). Thus, the sports press and historians were prompt to celebrate integration. In the 1980s, when the public debate on immigration first emerged, football was seen almost as a natural antidote to sudden outbreaks of racist events. Alongside "contemporary assumptions about the virtues of sport" (Gastaut, 2004), the

---

2 See book review in *Revue européenne des Migrations internationales*, 9(3), 1993: 266.

history of immigration focused mostly on Republican values linked to 'integration'. Italian immigrants in France have long understood that they had to become integrated. Although instances of rejection could not be ignored, they were usually seen as barriers to integration.

The context in the 1980s led to this decade becoming an important period for the revival of memories. With the end of mass immigration and the election of a Socialist government, the question of migration was once more a key issue in national politics and outdated humanist positions were quickly replaced by the principle of reality (Gastaut, 2004). The rise of the Front National and the Bicentennial celebrations for the French Revolution revived the debate on the French nation, in which historians and intellectuals also took part (Noiriel, 1988). The Republic congratulated itself on its capacity to smooth over social differences and origins, while the extreme right deplored the diminution of French identity. The ability to integrate is a key element in the Republican system of values.

In this situation, the revival of memories of Italian immigration between 1970 and 2000 was due, in very large part in our opinion, to an appreciation of the way in which Italians had succeeded in becoming integrated. At a time when North African immigration was seen as a problem for French society, many participants in this debate referred to the ability of Italian migrants to construct memory. For eulogists of the Republic, the integration of Italians was proof of the validity of the Republic's values and in particular of the role played by the national education system. On the other hand, those who saw North African immigration as a menace felt that Italian migrants' ability to integrate was proof that the French could only welcome 'cousins' with whom they shared close cultural and especially religious ties.

Memory and commemoration (2000s)

By the end of the 1990s, the context seemed to change. Although there was still a substantial debate on multiculturalism in France, this theme could easily be undermined whenever the concept of "a single and indivisible Republic" was challenged. At the same time, there was clearly a strong social need for memory, expressed by frequent references to a "duty to remember". According to historian François Dosse (1998), this soon led to "memory overkill". The philosopher Paul Ricœur (2000) was also troubled by the "worrying spectacle engendered by too much memory on one side, and too much forgetfulness on the other, even though little was said about the impact of commemorations or the abuse of memory and forgetfulness".

At the time, identity, memory and migration were clearly becoming interwoven. For Pierre Nora (2002), "identity, like memory, is a kind of duty. I am obliged to become what I am: Corsican, Jew, worker, Algerian, black, etc. It is this level of obligation that provides the link between memory and identity".

An editorial in the journal *Diaspora*, entitled "Migration and memory", went further: "Minority groups are knocking at the door and want to be heard and take their place [in society]. In their quest for recognition, some communities will go to unprecedented lengths to explore their past" (2005): Italians certainly fall into this category. In these circumstances, "memory sites" (a term coined by Pierre Nora) have multiplied, as have commemorations; indeed, they often coincide.

The commemoration phase also corresponds to the end of the migration process. Having become integrated, individuals with shared origins could peaceably reflect on their past and pay homage to those who made their presence in France possible. With the sense of belonging to the nation no longer an issue, the memory of Italian immigration could now be expressed through appropriate structures at the regional level, notably through associations.

An increasing number of associations were now acting as a stimulus for memory. The *Dante Alighieri Society* played an important role in this, thanks to its longevity, local networks and links with Italian state institutions (Salvetti, 1995). The *Dante Alighieri Society* was founded in Rome in 1889 with the objective of "defending and spreading the Italian language and culture throughout the world by reviving spiritual links with the homeland amongst compatriots living abroad and inspiring foreigners to love and admire Italian culture and civilization" (Caparelli, 1987, 278-279). Benefitting from support from the State, the Society now has 400 committees throughout the world, with 27 in France. Many of these committees are in frontier regions and, in particular, the South-East where there is a long and strong tradition of Italian migration, and most were created after 1945.[3] They often serve as a focal point for descendants of Italian migrants who find, through language classes, lectures and organized travel, a way to rediscover their origins. Indeed, many lectures address the question of Italian immigration and are given by historians.

In a very large number of Franco-Italian associations, the articles of association make a point of stating their desire to bring together people

3 In 2012, there were Dante Alighieri societies in Agen, Aix-en-Provence, Avignon, Bastia, Bordeaux, Cannes, Chambéry, Dijon, Draguignan, Le Mans, Lille, Limoges, Lyon, Metz, Modane, Montauban, Montpellier, Mulhouse, Nantes, Nice, Nîmes, Orléans, Paris, Perpignan, Toulon, Toulouse, Tours, Saint-Jean de Maurienne, and Sète.

of Italian origin (Mourlane, 2001). Usually, they develop a local base and encourage links with the local migrant community's region of origin. In the Paris area, these associations create networks via an Internet site.[4] Internet and new forms of communications have added a new dimension to the process of commemoration: membership of France's Italian community, via Internet, transcends – and enhances – local membership and other links (Graziano, 2012).

Other associations specifically seek to develop memory of origins and migration through genealogy. These associations are part of the French fascination for genealogy: it is estimated that amateur genealogists represent a large majority of public visitors to departmental archives (Mergnac, 2003). In this case, family memories can become confused with memories of immigration. We can cite one such association, *Ancêtres Italiens* (Italian Ancestors), which was founded in Paris in 1991 and offers 'personalized help' for people wanting to discover their Italian forebears.[5] Sometimes these associations organize cultural activities that give visibility to the memory of immigration. Where they are not themselves the initiators of such activities, they will become partners or help promote them.

They are also very active in twinning programmes. Twinning between French and Italian towns contributes to the establishment of transnational links between countries of destination and of origin and thus offer an original way to commemorate a migrant network. A study carried out in South-West France showed that links to the history of immigration are mentioned in many twinning programmes (Teulières, 2011b).

Since 2000, the number of exhibitions on Italian immigration has increased substantially and these are sometimes organized as part of a much larger programme on memory. For example, an exhibition organized in Paris in 2012 *Sur les traces de l'immigration italienne* ("Following the traces of Italian immigration") was the result of a project launched in 2010 to collect stories and songs from the Italian community living in East Paris. These exhibitions often have a local dimension, such as *Un air d'Italie* ("An Air of Italy") organized by the Musée Dauphinois in Grenoble from 18 November 2011 to 17 September 2012.[6] It is no coincidence that these exhibitions are being held in regions with a strong tradition of Italian migration. There are many Italians in Grenoble, where they represent 18% of the city's population in 1931 and 51% of all foreigners in 1958 (Vial, 2011). Its proximity to

---

4 http://www.associazioni-italiane.org/le-reseau (accessed 12.7.2012).

5 http://www.geneaita.org (accessed 13.7.2012).

6 http://unairditalie-museedauphinois.blogspot.it (accessed 13.7.2012).

the frontier gives greater visibility to the presence of Italians than at the national level. *A contrario* it should be noted however that, while the permanent exhibits at the Cité Nationale de l'Histoire de l'Immigration in Paris give prime position to Italian immigration, no temporary exhibition on this theme has been organized since this museum opened in 2007.[7]

Among the various cultural events relating to the memory of Italian immigration, live entertainment is becoming increasingly important. For example, as part of a recent event in the Paris suburb of Drancy on 14 April 2012, the association *Drancitalie*, founded in 2002, put on a show entitled *Italiens quand les émigrés c'était nous* ("Italians, When We Were the Emigrants"), an "exceptional spectacle of images, songs, and stories inspired by Italian emigration".[8]

This use of artistic events not only contributes to memory, but also takes on an educational, or even political, role in France where immigration has often been stigmatized as part of the debate on national identity launched by President Sarkozy (Noiriel, 2007). The same comment could be made of the play *Sale août* ("Dirty August") staged in 2010. Based on research by the historian Gérard Noiriel, this play was commissioned by the Cité Nationale de l'Histoire de l'Immigration in memory of a massacre that took place in Aigues-Mortes in 1893. In its review of the play, the daily newspaper, *Libération*, claimed that: "Pogrom, scandalous trial, stifled memory: the Aigues-Mortes massacre is not one of the most glorious pages in France's history".[9] Let us rapidly review the facts: on 17 August 1893, riots broke out between workers in the salt marshes of Aigues-Mortes and ended with 9 deaths and more than 50 wounded – the victims were Italians massacred by French people. All the assassins were acquitted and the affair was shelved (Barnaba, 1993; Noiriel, 2010). According to the programme notes, the play sought "to resurrect this forgotten and emblematic event that had been ignored for so long". But for *Libération*, it represented "contemporary drama", particularly because of the local population's passivity

---

7 The list of temporary exhibitions: *1931, les étrangers au temps de l'Exposition coloniale* ("1931, Strangers during the Colonial Exhibition", 6 May-7 September 2008); *À chacun ses étrangers* ("To Each Their Stranger", 16 December 2008-19 April 2009); *Générations, un siècle d'histoire culturelle des Maghrébins en France* ("Generations, a Century of Cultural History of North Africans in France", 7 November 2009-18 April 2010); *Allez la France! Football et immigration, histoires croisées* ("Go France! Football and Immigration, Crossed Histories", 25 May 2010-2 January 2011); *Polonia. Les Polonais en France* ("Polonia, Poles in France", 2 March 2011-28 August 2011).

8 http://drancitalie.pagesperso-orange.fr (accessed 13.7.2012).

9 René Solis "'Sale Août' au pogrom", *Libération*, 10 January 2011.

towards these events. Here, memory of Italian immigration serves, from the militant perspective, as a parable for current events.

According to Gérard Noiriel (2010), "the publication of new studies by researchers since 1993 has opened up a new phase in the collective memory of this massacre of Italian workers. In this process, the decisive moment was the centennial celebrations. The Société d'Histoire et d'Archéologie organized a seminar at Aigues-Mortes. A small commemorative plaque was placed on the path of historical walks allowing tourists to discover the rich past of the medieval town. (…) From now on, it is part of history because historians from major French and Italian universities have acknowledged its importance and awarded it the label of an 'historical event'".

It is important to note that commemoration of the events in Aigues-Mortes identified two key features: labour and sacrifice. In addition, these two features have long been the cornerstone of the memory of emigration in Italy which has always had a strong tendency to dwell on the dark side of events (Colucci, 2008).

Sacrifice and labour are also evoked when discussing events that took place in Izourt in the Pyrénées (Teulières, 2011a). In March 1939, there was another tragedy: labourers, mostly Italians, were working at high altitudes on the construction of a dam when a storm destroyed their campsite. As a result, 28 men died. For a long time this drama was forgotten but, 50 years later, the first commemorations were held in memory of the French rescuers. In 2002, an association was set up with the objective of erecting a memorial to the Italian victims. In this mountainous region, which was suffering from deindustrialization, the commemoration for the Italian victims formed part of the local population's desire to identify with their lost past. Memory of immigration had now merged, as at Aigues-Mortes, into local memory.

In our discussion of memory sites that are meaningful for the Italian immigration, let us turn to the neighbourhoods in larger French towns where Italians tended to congregate. Although the expression "Little Italies" is increasingly used to refer to these neighbourhoods in historical research – an epistemological approach in which memory clearly plays an important role – it certainly does not reflect the same reality as seen in the United States and, more specifically, the original "Little Italy" in New York (Blanc-Chaleard *et al.*, 2007). In France, Italian identity has never been so obviously linked to urban neighbourhoods and culture. In fact, there are no Italian ethnic neighbourhoods with a sufficiently strong sense of heritage that cities could, when developing tourist activities, erect memorials to the daily life of this migrant community.

The media is often a useful intermediary for the memory of Italian immigration. Historians have taken a growing interest in the immigrant press (Principe, 2009): for example, a magazine entitled *Radici* (*Roots*) which was founded in Toulouse in 2002 with a readership composed mainly of "past Italian emigrants, second and third generation descendants of emigrants, or people interested in Italy".[10] It focuses mainly on contemporary Italian culture and each month offers articles on the memory of immigration. It has strong links with the publishing house, Editalia, which also produces many works on this issue, including *Italians 150 ans d'émigrations en France et ailleurs* ("Italians, 150 years of emigration to France and elsewhere") which contained articles by historians and witnesses as part of the celebrations for 150 years of Italian unity. There is also *La Voce*, "the magazine for Italian identity and diversity in France" which was first published in 1998 and focuses on various themes relating to Italian cultural and social heritage.

Memory of Italian immigration nevertheless does not spark much interest in the broadcasting media. In the last years, only one film, *Bella Ciao* (2001), refers to Italian migration, and specifically to Italian Communists seeking to emigrate to the United States in 1932 but ending up in Marseilles.

We should also mention the Italian Film Festival held in Villerupt. Located in the mining region of Lorraine, Eastern France, Villerupt was home to a large number of Italian labourers. The film festival was first held in 1976 and is now a major cultural event. In the early years, most of the spectators were mainly of Italian origin: by watching these films, they were able to connect with their origins. The festival now attracts French audiences and, as a result, contributes to the restoration of the image of Italian migrant labourers and more generally of Italian culture. As it becomes increasingly well-organized, it promotes a positive image of Italians to the entire region and this reflects back on the Italian people themselves (Galorro and Boubeker, 2012). The Italian-Lorraine sculptor, Amilcar Zannoni, created the *Amilcar* trophy awarded to the festival's winners. For him, this event offers: "a cinema that is both committed and humane, and is enjoyed by a population proud to play such an important role in Lorraine's cultural history".[11]

The Villerupt Festival inspired one of the rare documentaries on Italian immigration *L'Anniversaire de Thomas: Villerupt, ville du fer* ("Thomas' Birthday: Villerupt, the Iron Town", 1982). In 2008, Marco Astolfi made a documentary film on the emigration of anti-Fascists to

[10] http://www.radici-press.net (accessed 13.7.2012).

[11] See Festival Brochure, 35th Festival of Italian Cinema, Villerupt (26.10-11.11.2011): http://www.luxbel.be/illustrations/villerupt2706.pdf (accessed 13.7.2012).

France, *Ciao Compagni, Salut Camarades* ("Ciao Compagni, Hi Comrades") based on personal stories and archival images. However, his documentaries are not shown on French television. Only Jean-Pierre Vedel's documentary entitled *Vie rêvée des Italiens du Gers* ("Dreams of Italians from Gers") has been produced and broadcast by French public television in 2009. It was based on the book by Yolande Magni, *Une histoire de promesse* ("Story of a Promise")[12] which gives a fictional account of an Italian community living in a small village in South-Western France. The book and the film focused on how France's Republican education system contributes to the integration and social mobility of Italians. No one has yet made a television documentary series on the history of Italian immigration in France.

Conclusion

These few references to the memory of Italian immigration in France since the 1960s, when the flow of trans-alpine migrants was beginning to wane, demonstrate the rich contribution this type of analysis can offer for understanding a migratory phenomenon occurring in different time periods. Without a doubt, several aspects merit greater attention. Historians of immigration and their current fascination for the question of memory are in fact opening up new lines of research and heuristic options.

In conclusion, we can now put forward a few hypotheses. First and foremost, the re-emergence of the memory of Italian immigration is part of a general nostalgia for roots. This leads to an increasing number of different types of events which construct memory, including, in particular, academic or artistic collection of oral history in which the two aspects are increasingly intertwined.

Next, memory is clearly produced within the context of the changing relationship between French society and Italian otherness. Italians are no longer excluded but rather are given legitimacy (Galloro, 2007) for reasons which go way beyond the long history of Italian migration and its good socio-economic integration in France. On the one hand, Italians conform to an image of model immigrants who easily accept the expectations of assimilation, an objective promoted by the state and applauded by public opinion, in sharp contrast to North African immigrants who are too obviously different. On the other hand, Italian migrants and their descendants benefit from representations that the French have of Italy, which has evolved from a marked condescendence to a more positive attitude since the 1970s (Milza, 1994). In addition, memory of Italian

12 Published by Elytis (Bordeaux) in 2009.

immigration contributes to the construction of transnational memory for the newly emerging European identity.

In this debate, frontier regions, where there is a long and strong tradition of Italian migration, play a fundamental role. As test areas for welcome or for rejection, as symbols of fears with regard to a diluted 'national identity', they are dynamic places of commemoration of the double identity, generation after generation. Memory of Italian immigration thus becomes a positive and separate element of local identities, particularly in Lorraine and the South-East.

Memory also focuses on integration as a positive factor, but it cannot sublimate the various forms of rejection, thanks to the contribution from historical research. Integration is no longer linear, as can be seen in part in the way that French public opinion of Italian migrants vacillates between acceptance and rejection. The dialectic relationship between history and memory is in fact central in the production of memory with regard to Italian immigration in France. Historians are thus both observers and actors.

References

2005. "Migrations en mémoires". *Diaspora*, No. 6.

Barnaba Enzo. 1993. *Le sang des Marais. Aigues-Mortes, 17 août 1893, une tragédie de l'immigration Italienne.* Marseille: Via Valeriano.

Bevilacqua Piero, Andreina De Clementi and Emilio Franzina (eds.). 2002. *Storia dell'emigrazione italiana.* Roma: Donzelli.

Blanc-Chaléard Marie-Claude. 2000. *Les Italiens dans l'Est parisien. Une histoire d'intégration.* Rome, École française de Rome.

Blanc-Chaléard Marie-Claude (ed.). 2003. *Les Italiens en France depuis 1945.* Rennes: Presses universitaires de Rennes.

Blanc-Chaléard Marie-Claude, Antonio Bechelloni, Bénédicte Descamps, Michel Dreyfus and Éric Vial (eds.). 2007. *Les Petites Italies dans le monde.* Rennes: Presses universitaires de Rennes.

Brierre Jean-Dominique. 2005. *Serge Reggiani. C'est moi, c'est l'Italien.* Paris: Hors Collection.

Caparelli Filippo. 1987. *La "Dante Alighieri".* Rome: Bonacci.

Colucci Michele. 2008. "Storia o memoria? L'emigrazione italiana tra ricerca storica, uso pubblico e valorizzazione culturale". *Studi Emigrazione*, 167: 721-28.

Colucci Michele and Matteo Sanfilippo. 2010. *Guida allo studio dell'emigrazione italiana.* Viterbo: édizioni Sette Città.

Corti Paola. 2003. "L'emigrazione italiana in Francia: un fenomeno di lunga durata". *Altreitalie* 26: 4-24.

Corti Paola and Matteo Sanfilippo (eds.), 2009. *Storia d'Italia.* Annali 24: *Migrazioni*. Torino: Einaudi.

Dornel Laurent. 2004. *La France hostile. Socio-histoire de la xénophobie en France (1870-1914).* Paris: Hachette-Littératures.

Dosse François. 1998. "Entre histoire et mémoire: une histoire sociale de la mémoire". *Raison présente?*: 5-24.

Faidutti-Rudolph Anne-Marie. 1964. *L'immigration italienne dans le Sud-Est de la France.* Gap: Ophrys.

Galloro Piero. 2007. "La 'trace' des Italiens établis en Lorraine: de l'exclusion à la legitimation". in Boggiani J., M-L. Caldognetto, C. Cicotti, and A. Reuter (eds.) *Traces de mémoire, mémoire des traces – Parcours et souvenirs de la présence italienne au Luxembourg et dans la Grande Région*, Luxembourg/Esch-sur-Alzette: Presses Universitaires Luxembourg: 91-118.

Galloro Piero and Ahmed Boubeker. 2012. *L'immigration en héritage – L'histoire, la mémoire, l'oubli aux frontières du grand Nord-Est.* Nancy: Presses Universitaires de Nancy.

Gastaut Yvan. 1996. "Les tendances italophobes dans l'opinion niçoise à la Libération". *Cahiers de la Méditerranée*, 52: 33-57.

Gastaut Yvan. 2000. *L'immigration et l'opinion en France sous la Ve République.* Paris: Seuil.

Gastaut Yvan. 2004. "Français et immigrés à l'épreuve de la crise (1973-1995)". *Vingtième Siècle* 84: 107-118

Gastaut Yvan. 2008. *Le métissage par le foot, l'intégration mais jusqu'où.* Paris: Autrement.

Gastaut Yvan and Stéphane Mourlane. 2009. "Le 'rital' et le 'nord-africain' ou la pertinence du bouc-émissaire". Amar Marianne, Marie Poisot, and Catherine Wiithol de Wenden (eds.) *A chacun ses étrangers? France-Allemagne de 1871 à aujourd'hui.* Arles/Paris: Actes Sud/CNHI: 93-99.

Girard André and Jean Stoetzel. 1954. *Français et immigrés. Nouveaux documents sur l'adaptation*, Paris: PUF, Cahier de l'INED 20.

Graziano Teresa. 2012. "The Italian e-Diaspora: Patterns and practices of the Web". *e-Diasporas Atlas.*

Mergnac Marie-Odile (ed.). 2003. *La généalogie. Une passion française.* Paris: Autrement.

Milza Pierre. 1981. *Français et Italiens à la fin du XIXe siècle.* Rome: École française de Rome.

Milza Pierre. 1993. *Voyage en Ritalie*. Paris: Plon.

Milza Pierre. 1994. "L'image de l'Italie et des Italiens du XIXe siècle à nos jours". *Cahiers de l'IHTP* 28: 71-82.

Milza Pierre and Marie-Claude Blanc-Chaléard. 1995. *Le Nogent des Italiens.* Paris: Autrement, Français d'Ailleurs, Peuples d'Ici.

Mourlane Stéphane. 2001. "Solidarités formelles et informelles: les associations d'Italiens en France depuis 1945". *Cahiers de la Méditerranée* 63: 199-211.

Mourlane Stéphane. 2002. *Une certaine idée de l'Italie. Attitudes et politique françaises 1958-1969.* PhD thesis in History, University of Nice.

Mourlane Stéphane. 2003. "Michel Platini et l'Italie: les origines en question". *Migrance*, 2: 111-118.

Mourlane Stéphane. 2007. "Que reste-t-il des préjugés? L'opinion française et l'immigration Italianne dans les années 1950-1960". *Migrations société*, 19(109): 133-145.

Mourlane Stéphane. 2010. "Piantoni et Platini, ces "héros Italians" du football français". Boli Claude, Yvan Gastaut, and Fabrice Grognet (eds.), *Allez la France! Football et immigration*. Paris: Gallimard/CNHI/ Musée national du Sport: 143-145.

Noiriel Gérard. 1988. *Le creuset français. Histoire de l'immigration XIXe-XXe siècle*, Paris: Seuil, L'Univers Historique.

Noiriel Gérard. 2001. *État, nation et immigration. Vers une histoire du pouvoir.* Paris: Belin.

Noiriel Gérard. 2005. "Les chantiers de l'histoire. Historiographie de l'immigration". *Hommes & Migrations*, 1255.

Noiriel Gérard. 2007. *À quoi sert 'l'identité nationale'?* Marseille: Agone.

Noiriel Gérard. 2007. *Immigration, antisémitisme et racisme en France XIXe-XXe siècle. Discours publics, humiliations privées.* Paris: Fayard.

Noiriel Gérard. 2010. *Le massacre des Italiens – Aigues-Mortes, 17 août 1893.* Paris: Fayard.

Nora Pierre (ed.). 1984-1993. *Les Lieux de mémoire*. Paris: Gallimard.

Nora Pierre. 2002. Pour une histoire au second degré. *Le Débat*, 122.

Panicacci Jean-Louis. 2010. *L'occupation italienne. Sud-Est de la France, juin 1940-septembre 1943*. Rennes: Presses Universitaires de Rennes.

Pasquini Pierre. 1993. "Le parler des immigrés: l'exemple des Italiens en Provence". Sanguin André-Louis (ed.), *Les minorités ethniques en Europe*. Paris: L'Harmattan: 149-154.

Principe Lorenzo (ed.). 2009. "La stampa di emigrazione italiana". *Studi emigrazione*, 175.

Ricœur Paul. 2000. *La mémoire, l'histoire, l'oubli*. Paris: Seuil.

Rygiel Philippe. 2009. "Archives et historiographie de l'immigration", *Migrance*, 33: 50-59.

Salvetti Patrizia. 1995. *Immagine nazionale ed emigrazione nella Società 'Dante Alighieri'.* Roma: Bonacci editore.

Sanna Giuseppina. 2011. *Il riscatto dei lavoratori. Storia dell'emigrazione italiana nel sud-est francese (1880-1914).* Rome: Ediesse.

Sayad Abdelmalek. 1977. "Les trois âges de l'émigration algérienne en France". *Actes de la recherche en sciences sociales*, 15: 59-79.

Schnapper Dominique. 1974. "Centralisme et fédéralisme culturels: les émigrés italiens en France et aux États-Unis". *Annales ESC*, 5: 1140-1160.

Schor Ralph. 1985. *L'opinion française et les étrangers en France, 1919-1939.* Paris: Publications de la Sorbonne.

Schor Ralph, Stéphane Mourlan and Yvan Gastaut. 2010. *Nice cosmopolite, 1860-2010.* Paris: Autrement.

Tafuieff Pierre-André. 1988. *La force du préjugé. Essai sur le racisme et ses doubles.* Paris: La Découverte.

Tapinos Georges. 1975. *L'Immigration étrangère en France, 1946-1973.* Paris: PUF.

Temime Emile. 1985. "Marseille, ville de migrations". *Vingtième siècle. Revue d'histoire* 7: 37-50.

Temime Émile (ed.). 1989-91. *Migrance. Histoire des migrations à Marseille.* Aix-en-Provence: Edisud.

Teulières Laure. 2011a. "Barrage d'Izourt en Haute Ariège. Des ouvriers ont payé ici de leur vie". Teulières Laure (ed.), *Italiens. 150 ans d'émigration en France et ailleurs 1861-2011.* Toulouse: Éditalie: 418-425.

Teulières Laure. 2011b. "Le lien migratoire à l'honneur: les jumelages franco-italiens". Teulières Laure (ed.), *Italiens. 150 ans d'émigration en France et ailleurs 1861-2011.* Toulouse: Éditalie: 434-438.

Tribalat Michèle. 1991. *Cent ans d'immigration, étrangers d'hier, Français d'aujourd'hui. Apport démographique, dynamique familiale et économique de l'immigration étrangère.* Paris: Presses universitaires de France, Institut national d'études démographiques.

Vegliante Jean-Charles. 1995. "Représentations, expressions (un aperçu d'ensemble sur la culture Italienne immigrée en France)", in A. Bechelloni, M. Dreyfus and P. Milza (eds.), *L'intégration Italienne en France.* Bruxelles: Complexe.

Vial Éric. 2011. "Grenoble, ville italienne?" in Teulières Laure (ed.), *Italiens. 150 ans d'émigration en France et ailleurs 1861-201.* Toulouse: Éditalie, 122-130.

New Geographic Mobilities on the Edges of Greece

Pierre SINTÈS

This article is the result of various research studies I have carried out in Greece and the Balkans on the interconnectedness between the development of new geographic mobility and the reactivation of local memories and memories of minorities. This research has led me to develop working hypothesis that challenge the notion of social cohesion and fragmentation as part of contemporary world evolutions, and more specifically the shaping of local or transnational identities as a result of developments of transport and communications and their impact on actors' relationships with territories. Based on the observation of different areas (border regions, tourist areas, cities, diaspora networks), I am continuing my study of fragmentation processes that affect Greek society by evaluating the role of exogenous forces strongly linked to the enforcement of European policies and local dynamics which are often associated with questions of resilience and memory. This research is part of an on-going project initiated several years ago with a new research team on the contemporary Balkan region by opening up new fields of knowledge about the social situation in this area.

Research hypotheses and theoretical framework

My hypotheses stem from previous observations of international migration in Greece which are useful for analysing Greek attitudes to otherness and change. This new research has allowed me to widen the scope of my research to include the social impact of globalization on Greek society, by using as a starting point the issues raised by this phenomenon in terms of memory and links with territory.

Migration from Albania to Greece: a starting point

Studying migration from Albania to Greece in recent years has enabled me to explore a number of questions about contemporary Greece, and specifically on economic, social and political issues (Sintès, 2010a). Yet, in addition to a traditional study of the geography of migrations, my field research into how migration actually takes place has shown that the country has been forced to reconsider to its current and past situation against the larger backcloth of Balkan history. As a matter of fact,

Albanian migration to Greece has led to the reactivation of representations and of ancient historical links that unite the two countries and their inhabitants. A study of biographical narratives of Albanian migrants has revealed the value of understanding where individual migrants come from, where they are going, and the obstacles that lie in their path (Sintès, 2007); above all, it has thrown light on the importance of representations as part of the history shared by the two countries (Sintès, 2010a). The fact that migrants are part of a process of self-identification on different levels, as can be seen in other Balkan countries (Stahl, 1977; Gossiaux, 2002), puts them in an implicit position towards Hellenism. Because of specific historical conditions that resulted from the establishment of the border between the two countries, migrants are *de facto* divided into different groups: some migrants are considered to be members of the Greek nation in their own right, while others are seen as allies or enemies (Triandafyllidou & Veïkou, 2002). In Greece, membership of the first group grants migrants an administrative status that is very different from that of the other group: they are treated kindly by the local population, and even benefit from some forms of "affirmative action" in the labour market when work is scarce. The reverse is true for migrants in the other category which includes Albanian Muslims who suffer from a very negative image due to stereotypes widely promoted in the media (Nestoropoulou, 2008) or by the school system (Fragonaki & Dragona, 1997). These initial conclusions form the basis for my research into the interconnectedness with mechanisms of identification, a sense of belonging and the places experienced by migrants.

Globalization, mobility and fragmentation

Building on these first results, it seems to me that the development of international mobility has played an important role in these processes. In fact, it appears that the type of migrations that took place during the 19th and 20th centuries have given way to a polymorphous spatial circulation typified by Albanian migration in Greece (Sintès, 2007). These new forms of mobility, because they are reversible, may be seen as turning migrants into "circulating populations" that interact with the various places they visit, but no longer seek to settle permanently (Knafou, 2004; Retaillé, 2006). The fact is that today it has become much easier for migrants (though, of course, in varying degrees) to move around in the same way as goods and information: this phenomenon, which began during the industrial era (Grataloup, 2007) and has accelerated with the revolution in transportation and communication since the beginning of the 21st century, lies at the heart of changes in the way people relate to space. On a social level, these transformations entail a redefinition of the processes of integration and exclusion as discussed in the examples below. Globalization seems to bring a certain form of uniformization, a standardization of representations and practices, through the circulation of information and multiplication of exchanges. As such, it has an inherent movement of integration, partly because it asserts a hegemonic system of values, that Michael Herzfeld calls "the global hierarchy of values" (Herzfeld, 2004). And yet, it can be observed that this process goes hand-in-hand with its opposite, the apparition of new lines of fractures, which are becoming increasingly numerous, on all levels as a result of processes of fragmentation. These transformations in interaction between the centre and the periphery can also be seen to bring tremendous changes in the production of marginality and exclusion. Liminality no longer exists in terms of proximity to the centre. Topographic distances are less and less pertinent for explaining exclusion; they have been effaced by the need to have good connections with social and economic resources. In this new configuration, marginality is not just a fact of being on the periphery, it can appear anywhere in society, even in the interstices at its centre.

This reflection, which draws on urban geography research into the fragmentation of space in cities, can be applied to social mechanisms. It is now certain that identification with diasporas (in the context of migration), regional and local groups (in a context of sedentarism) is becoming increasingly important (while the standardization of lifestyles has never been more powerful) with the emergence of groups claiming to stand at the margin of (all levels of) societies. This phenomenon is growing but, instead of describing it as fragmented, I think the metaphor of the fractal often seems more appropriate. This mathematical image – which evokes a geometrical object the limits of which cannot be defined since the closer one looks at them, the more it appears to have an infinite number of facets, is often used to describe the Balkans. Sarah Green (2005) claims, for example, that the closer you come to the Epirus area that she studies, the more micro-fractures you find, i.e. micro groups that seem to be elements of infinite fragmentation of these societies: she calls this "the Balkans' fractal". Rather than this approach, I prefer Stjepan Mestrovic's analysis (1994), which does not focus specifically on regional issues and is more radical: he suggests that behind this fractalization, lies the process of globalization which, according to him, is taking place not just in the Balkans, but also throughout the whole world, from Los Angeles to Mumbai, via Sarajevo. From this perspective, social aspects of globalization are unfolding in the Balkans as elsewhere and leading to a fragmentation of society and spaces. In fact, it is often found in what Arjun Appadurai (2005) calls "primordialism", i.e. the mobilization of identity discourses inspired by pre-modern groups, which were often created before, and on a smaller scale than, the nation-state. This movement fits the definition of Jean-Loup Amselle's concept of "Retrorevolution" (Amselle, 2010). It echoes the political impact of globalization, such as changes in the balance of power that regulates space and the emergence of new actors (also made possible by geographic mobilities) claiming to belong to supranational or transnational identities. These changes may be identified as the cause of the troubles, the "turbulence" (to use James N. Rosenau's term), in the late 20th century and early 21st century, because they created an imbalance between traditional structures of power centred on nations, on the one hand, and new transnational power networks, on the other. Even in most recent research (Sassen, 2009; Abélès, 2008), social sciences make us focus more specifically on the framework and actors of these transformations in the role of nations, by taking into account questions of power and territories and thus returning to one of the key themes of geopolitics (Rosière, 2007). Recent events in Greece illustrate the radical forms that rivalry between different levels of power has taken in the outer edge of Europe, the Balkans.

Migration, memory and society in Greece

For several years, my research has followed this trajectory in Greece. From now on, I will focus on different situations which allow me to challenge these hypotheses. For each of the case studies, I want to observe international mobility in order to analyse how society reacts to the globalized world and above all to the changes in the hierarchy of actors generated by globalization. In Greece, my observations mainly occur in two specific areas of the periphery.

North-west Greece and its frontier with Albania

The first of these points of observation is located in Epirus, a region in North-west Greece. My research has led me to review the relationship between Albania and Greece by studying the controversial question of the Albanian-speaking community in Greece. Thus, I can question the interconnectedness between national narrative, ethno-linguistic identity and religious practices in a global context of social re-composition and migration.

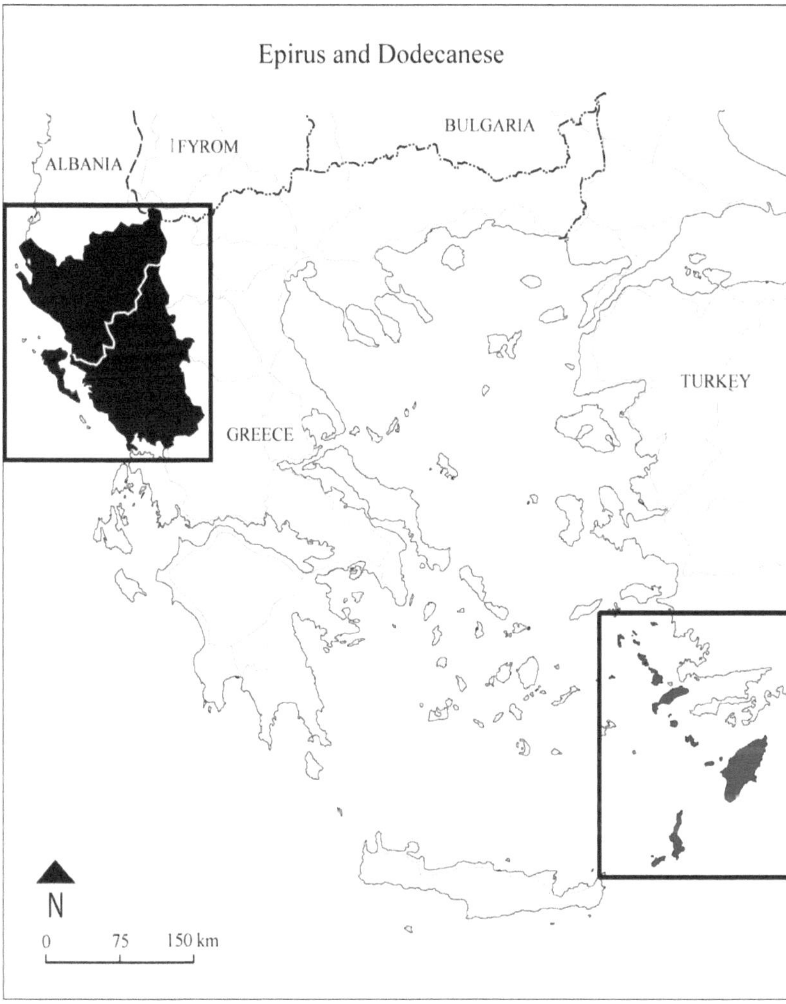

P. Sintès, 2012

Locally, these areas are very interesting because the way people settled in this region has changed tremendously since the Second World War; whole villages of Albanian-speaking Muslims were deserted and became "ghost villages". Their lands were redistributed to rural communities of former tenant farmers and to groups of people who had previously lived in the mountains and were relocated to new villages after the Greek Civil War. These stories of spoils and violence, which are still remembered today, have been revived by the arrival of Albanian migrant workers who were already entering Greek territory in large numbers well before they were no longer required to obtain visas after 15 December 2010. West of the Greek-Albanian border, the presence of **this socio-historical substrate** gives a particular dimension to on-going migration. As observed in more eastern areas of this borderland (Sintès, 2003; 2008; 2010a), after the frontier was reopened in the early 1990s, some inhabitants of Epirus, descendants of semi-nomadic Aromanian shepherds (Gossiaux, 2002), were able to reunite with relatives or friends who had stayed in Albania after 1945 when the border was closed. When these Albanians decided to migrate to Greece, they reactivated these family ties. But an even more novel phenomenon was the way that both "frontier life" and migration were facilitated, after 1990, by the previous coexistence of Christians and Muslims and of Albanian-speaking and Greek-speaking people dating back to the pre-1914 period. For example, in the most western part of the border, a tradition of good relationships between the Greek village of Sagiada and the Albanian village of Konispol was revived in a different way during the 1990s: when it became possible to migrate but the Greek police prevented Albanians from crossing the border, the inhabitants of Sagiada intervened in favour of their Konispol neighbours by stressing the traditional links between the two villages and their customary friendship. The revival of these transnational links also takes on a more personal dimension: the oldest inhabitants knew each other before the war, despite their ethnic differences, and were able to visit each other after the fall of the Enver Hoxha regime in 1990. This type of "memorial border crossing" also takes place among Albanians with no previous contact: people born in this region and displaced in 1945 have visited these "ghost villages". In Epirus the relationships created by these specific forms of mobility are facilitated by the older generations' ability to speak several languages (Kretsi, 2002) even though we can draw no conclusions on the inhabitants' self-identification from this fact.

This situation opens up a new perspective on the links between Greek and Albanian citizens, and thus suggests several paradoxes which oppose the villagers' sense of belonging and the national regulatory framework. Solving these paradoxes is often made possible by infra-national identities, identification with a small homeland or a network of solidarity. Actors sometimes feel this tension and, by carrying out a multilingual survey in these peripheral areas, it is possible to observe how the respondents handled the polyphony of their identity within this new situation, which is very similar to what has been observed with the second generation of immigrants in Greece (Gogonas, 2010). These transnational connections do not fit into the traditional frame of national identity, according to the people we met; rather, it is evidence of the existence of an "in-between" state that is not lacking in vitality. Using local identities is a handy way of overcoming difficulties and resolving what initially appeared to be a paradox. More than just a simple case of areas close to the border, it is impossible not to discuss the situation of all the Albanian-speaking population of Greece and their descendants, whose presence can be found in census records for the 1950s; their position and attitudes have undoubtedly changed because of the presence of the contemporary Albanian migrants (Sintès, 2010b). On a larger scale, the future of all multilingual populations in Greece can be questioned because of the development of international mobilities from the Balkans to Greece.

Albanian-speaking inhabitants in Greece, according to the censuses of 1920, 1928 and 1951. Number of Albanian speakers according to the Greek National Statistics Institute

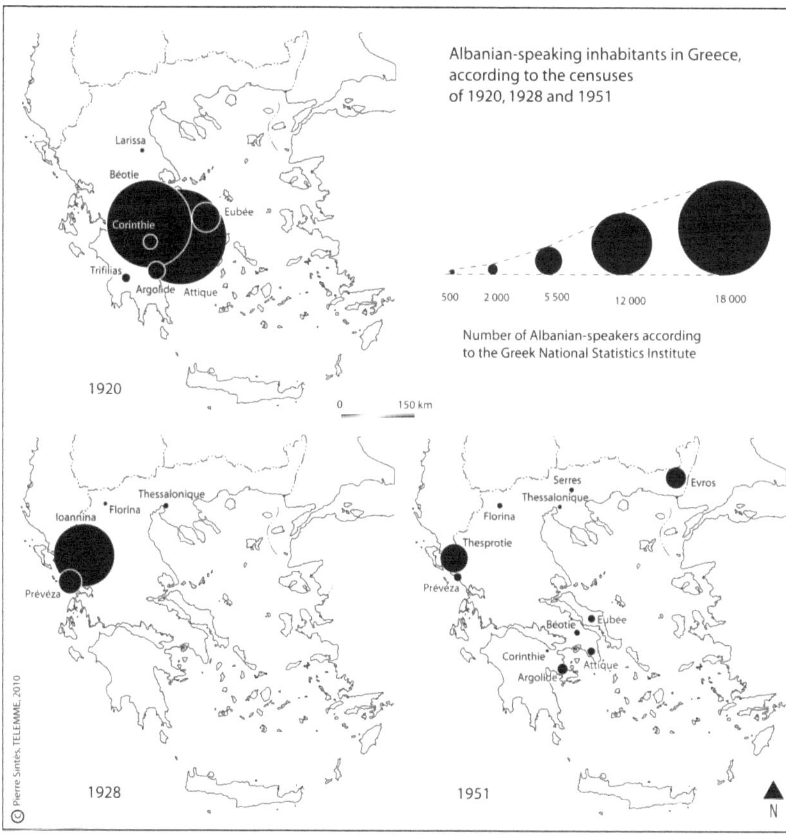

Tourist and memorial mobility, labour migration in Rhodes

The second study deals with a seemingly very different geographical context: the old city of Rhodes, capital of the Dodecanese archipelago. This tourist area is today at the heart of different forms of mobility and international migration which have affected its social functioning and its territories by reviving the island's history in an original way. I have studied this phenomenon for several years by looking into the arrival of descendants of the city's ancient Jewish community who have travelled to Rhodes for memorial visits, what is sometimes called "tourism of identity" or "root tourism" (Fourcade, 2010). This phenomenon throws new light on a particular aspect of globalization, since their presence is largely due to contemporary conditions for the mobility of people and information. In Greece, similar events occur in other areas involving descendants of groups that have since disappeared: other Jewish communities but also Muslim communities that were exchanged with Turkey in 1923 (Bayindir Goularas, 2010). In Rhodes, however, these tourists visit a place where the Jewish community has totally disappeared due to their massive emigration in the inter-war years and above all to the deportation of large numbers of Jews by the German army during the summer of 1944 (Fintz-Menascé, 1991). Over the last decade, we have seen a new and surprizing vitality in the public expression of this memory: the refurbishing of the oldest synagogue in the *Djuderia* in the heart of the medieval city; the opening of a museum, and in particular the annual commemoration events in memory of the deportees, which takes place on the main plaza of the former Jewish neighbourhood (Sintés, 2010c). These places of memory have become, for the descendants of Rhodian Jews, a focus for family pilgrimages and particularly during Jewish high holidays. During those trips, the relationships established with the local population interact with the difficult question of the preservation of the medieval city, which is now included in UNESCO's World Heritage List in 1988, and from which the former Jewish neighbourhood, as well as the traces of this very particular history, are clearly differentiated. Here, the question of transmission of houses which belonged to Jews who had been departed to their heirs is clearly posed. More generally, because the national narrative in Greece does not include either the Jewish or the Muslim community, this heritage is not taken into account in the urban planning, tourism and heritage programmes for this part of the medieval city in Rhodes (Sintès, 2011b).

This research, together with a broader study on Diaspora networks in this community, has led me to adopt a more global approach to the situation in Rhodes which ties in more closely with my project to analyse fragmentation processes in contemporary Greek society. In fact, as in Epirus, the dynamics at stake in the old city as a result of new international mobilities (tourists, memorial or labour migrations) have also challenged the conventional discourse on the Greek Nation-State within its territory. The Rhodian context features a new set of international or transnational "transgressive" actors for just such a purpose: the Jewish community, Turkish-speaking Muslims and Catholics, all of whom have a historical presence on the island, together with a very large number of labour migrants (mainly Albanians, Bulgarians and Filipinos) since the 1990s. They are all key actors in the Rhodian situation and the importance of their contribution to the island's attractiveness for tourists is clearly expressed by other local actors (Sintès, 2011b). The conflict between the discourse of the nation-state and the social and economic reality allows us to see that, in Rhodes as in our first case study, the question of the revival of an identity that is specific to a peripheral group to the border area or a minority can be caused by recent developments of geographic mobility. As elsewhere, changes in hierarchy because of these migratory movements or international mobilities, on the one hand, and the sudden emergence of social actors such as UNESCO or the EU, on the other, provides another opportunity for understanding how the globalization of processes and fragmentation phenomena occur simultaneously in these border spaces because of mobilities and migrations.

International migration in the poorer districts of North-West Athens

These on-going studies in Rhodes and Epirus can be enriched by opening up another opportunity for observation, this time in an urban setting: Athens. This stems from a need to understand the new dynamics of marginalization and fragmentation, with regard to the current situation in the country due to the 2008 financial crisis, the effects of which were most keenly felt in urban areas. For this study, a number of places could be selected for observation and comparison with the geographical peripheral areas in Epirus and the Dodecanese: the western suburbs of Athens, Peristeri and Aigaleo which were created to accommodate a large number of refugees from Asia Minor in 1923, or poorly integrated areas in the city centre, such as Patissia, where there were violent outbursts in 2009 sparked by the presence of many foreigners. These districts, which have been badly hit by the economic and political crisis, have highly visible populations of migrants due to the settlement of many Albanians in the 1990s and migrants from more remote countries in the first decade of 2000, mostly from Central Asia and West Africa (Sintès, 2011a). These poorer districts have become places of often violent confrontations between anti-racist groups and new "neighbourhood self-protection groups", broadly structured by the extreme right party *Chrisi Avgui* (Golden Dawn), over the presence of these migrants. They offer a classic example of fragmentation in an urban space that was once quite homogeneous (Sintès, 2002; Maloutas, 1992) but where there is now a concentration of migrants and of the poorest nationals.

Yet, quite apart from these tensions and confrontations, the specific past of these districts and their relationship with the history of forced mobilities during the 1920s and 1940s, have become an important element in terms of the municipal authorities' political and identity discourse. They have thus attracted public attention to a tradition of hospitality and loyalty and their attachment to the values of Hellenism (Parsanoglou, 2010), which sets them apart. Focus on this situation will provide us with a new point of comparison in terms of the relationship between past and present which is having a growing influence on current attitudes and practices in Greece. It will also enable us to show how this relationship is being (re)composed in the context of an acute crisis that affects the whole of Greek society.

Conclusion

The comparison of these different areas in Greece allows us to pinpoint the tensions that have affected Greek society over the last decade and the relationships that link memory, international mobility and the emergence of marginality. While revealing stark contrasts which nevertheless reflect the many facets of a country, these case studies offer an efficient analysis of how contemporary geographic mobilities can play a key role as factors of social or spatial fragmentation, while also ensuring that Greece is an integral part of our world. The creation of transnational ties and strengthening of local identities are also encouraged through discourses that promote a particular approach to the handling of memory: in Rhodes, conservation of a past heritage, in Epirus, respect for linguistic and religious minorities, and in Athens, the historic specificity of poorer western suburbs. The intervention of supernational actors, including international and European bodies whose role in recommending standards has been analysed, among others, by Sabine Riedel's studies in Romania and Hungary (Riedel, 2002), contributes, here as elsewhere, to this new situation. More globally, by reviving minority or marginal memories and by the presence of migrants and other mobile groups, we can discuss the weakening, or rather the displacement, of the nation-state as part of our current world.

In this article, I do insist on the importance of tackling these social transformations by concentrating on the territorial angle, i.e. the production of places or networks of belonging. These thematic exchanges allow us to develop an approach that will highlight the evolution of contemporary Balkan societies with a research team that we are now putting together. More particularly, the development (or the transformation) of the role played by the European Union must be stressed as a process that is emblematic of globalization mechanisms as they impact Balkan countries. The fact that violence has ended in the Balkans should not allow us to forget what is happening as a result of the construction of Europe and the redefinition of European frontiers. The integration of Bulgaria and Romania on 1 January 2007 and the association and stabilization agreements signed with Western Balkan countries raise a number of questions about the future of these marginal European regions, but also about the European project itself: what does European integration mean for the inhabitants and countries in these regions? European integration is often evoked to justify brutal social and economic transformations. It is sometimes accused of being the cause of all the difficulties that the Balkans peoples must face, such as financial crises and inflation. To what extent does it offer new opportunities or, on the contrary, involve new limitations to which individuals must react and adapt? The European Union is also an important prescriber of norms and plays an essential social role. How are categories created in the West (such as 'heritage') understood and reinterpreted locally? The Declaration of Independence by the Province of Kosovo on 17 February 2008 also demonstrates that political re-composition can still cause tensions and bring transformations in the definition of borders and the sense of belonging. Besides, the Balkan countries have become the door into Europe for migrants from the East and South, as has been shown by recent works on the circulation of Afghans (Tarrius, 2007). Therefore it is legitimate to think that, in the years to come, this region will play a key role in issues relating to mobilities and border controls at the national and European level (cf. the article by Pillant in this volume).

Questioning the "Europeanization" of the Balkans (Anastasakis, 2005) could take advantage of this analysis. The configurations that are revealed, albeit partial, allow us to have a clear view of what is at stake. Migration and mobility are key topics, since the various types of geographical movements, and their control, are main vectors for the new norms being imposed by the European Union. Besides, mobility and migration seem to be important elements in the redefinition of territories because they give places different functions depending on the perspective chosen, whether of researchers, inhabitants or administrators.

The same can be said about heritage and the status of memory, topics which allow us to tackle the issue of transformation of time and its entrenchment in the territory, by studying aspects to which the dominant discourse allocates value and for which respect for authenticity and the regenerative values of the past are inherent. This new dimension takes us beyond the simple nation-state to the international valorization of heritage. Undeniably sustained by the EU's policy for integration, this discourse provides legitimacy to different new institutional and social actors (UNESCO, NGOs, diaspora). It can also contribute to the process of fragmentation of the society, now being observed in the peripheral areas of Greece (Sintès, 2011b). Finally, the status of interstate borders allows us to highlight the social transformations produced by EU policies in these specific territories. Indeed, this lies at the heart of interconnections between the different levels of powers and of legitimacies in the administration of territory.

A methodology consisting of a meticulous study of concrete cases provides an opportunity to analyse the complexity of stratified logics relating to these phenomena and "how such entities as locality, community and region are constructed and understood" (Harvey, 1993; 6). Examining these local configurations also helps us to understand changes in the sense of belonging and in actions that play an important part in the way places, territories and identities are displayed. This type of analysis also allows us to examine the processes of discrete resistance, which become invisible when observed from a larger perspective, in order to show how actors invent ways to make use of, or undermine, dominant ideologies by positioning themselves between the local area and the global world (Abélès, 2008). Finally, these case studies allow for a multi-sited observation of the 'fractality' that defines Balkan countries, according to Sarah Green (Green, 2005).

References

Abeles M. 2008. *Anthropologie de la globalisation*, Paris: Payot.

Amselle J.-L. 2010. *Rétrovolutions*, Paris: Stock 'Un ordre d'idées'.

Anastasakis O. 2005. "The Europeanization of the Balkans". *The Brown Journal of world Affair*, Vol. XII No. 1/ 77-88.

Appadurai A. 2005. *Après le colonialisme. Les conséquences culturelles de la globalisation*. Paris: Payot & Rivages.

Bayindir Goularas G. 2010. "Vers les terres natales, vers les terres des ancêtres". *Téoros, Revue francophone de recherche en tourisme*, "Le tourisme des Racines". Quebec: Presses universitaires du Québec: 3-6.

Davis M. 1990. *City of quartz: Excavating the Future of Los Angeles*. London: Verso.

Fintz-Menascé E. 1991. *Gli ebrei a Rodi*. Milan: Guerini.

Fourcade M.-B. 2010. "Tourisme des racines". *Téoros*, *Revue francophone de recherche en tourisme*, numéro spécial "Le tourisme des Racines", Quebec: Presses universitaires du Québec: 3-6.

Fragoudaki A. and T. Dragona (eds.). 1997. *Τί είν' η πατρίδα μας*. Athens: Alexandria.

Gogonas N. 2010. *Bilingualism and Multiculturalism in Greek Education*. Cambridge: Cambridge Scholars Publishing.

Gossiaux J.-F. 2002. *Pouvoirs ethniques dans les Balkans*. Paris: Presses universitaires de France.

Grataloup C. 2007. *Géohistoire de la mondialisation. Le temps long du Monde*. Paris: Colin, collection U.

Green S. 2005. *Notes from the Balkans. Locating Marginality and Ambiguity on the Greek-Albanian Border*. Princeton: Princeton University Press.

Harvey D. 1993. *From Space to Place and Back Again: Reflections on the Condition of Post-Modernity*. New York: Routledge.

Herzfeld M. 2004. The *Body Impolitic: Artisans and Artifice in the Global Hierarchy of Value*. Chicago: University of Chicago Press.

Kretsi G. 2002. "The 'secret' of the past of the Greek-Albanian borderlands. Cham Muslim Albanians: Perspectives on a conflict over historical accountability and current right". *Ethologia Balkanica*, 6: 171-195.

Knafou R. (ed.). 2004. *La planète "nomade". les mobilités géographiques aujourd'hui*. Paris: Belin.

Maloutas T. 1992. "Dissociation sociale à Athènes", in T. Maloutas and D. Economou (eds.), *Structure sociale et organisation urbaine à Athènes* (in Greek), Athens: EKKE: 67-140.

Mestrović S. 1994. *The Balkanization of the West: The Confluence of Postmodernism and Postcommunism*. London: Routledge.

Nestoropoulou A 2008. "L'altérité dans la presse régionale, l'exemple du département d'Imathias", in G. de Rapper and P. Sintes (eds.), *Nommer et classer dans les Balkans*. Athens: École française d'Athènes (Mondes Méditerranéens et Balkaniques): 169-187.

Parsanoglou D. 2010. *Grèce, pays d'immigration: Perspectives historiques et sociologiques*. PhD thesis, EHESS.

Rétaillé D. 2006. "La géographie de demain et d'ailleurs". Brouillons Dupont, Conference papers Géopoint *Demain la Géographie, pourquoi? comment?* Université d'Avignon.

Riedel S. 2002. "Minorités nationales et protection des droits de l'homme, un enjeu pour l'élargissement". *Politique étrangère* 3: 647-664.

Rosenau J. 1992. *Turbulence in World Politics*. Princeton: Princeton University Press.

Rosière S. 2007. "Comprendre l'espace politique". *L'espace politique* 1: 1 | 2007-1, placed on line 16.1.2007, http://espacepolitique.revues.org/index 223.html (accessed 28.9.2012).

Sassen S. 2009. *La Globalisation, une sociologie*. Paris: Gallimard.

Sintès P. 2002. "Immigration, réseaux et espace métropolitain, le cas athénien". *Les Cahiers de la Méditerranée*, 63: 311-329.

Sintès P. 2003. "Les Albanais en Grèce: le rôle des réseaux préexistants". *Balkanologie* VII(1): 111-133.

Sintès P. 2007. "Les mobilités des travailleurs balkaniques en Grèce: migration de travailleurs ou circulation régionale?". *L'Espace géographique*, No. 36-3, Paris: Belin: 353-365.

Sintès P. 2008. "Les Valaques de la région de Gjirokastër et la Grèce". In G. de Rapper and P. Sintes (eds.) *Nommer et classer dans les Balkans*. Athens: École française d'Athènes, Mondes Méditerranéens et Balkaniques: 23-43.

Sintès P. 2010a. *La raison du mouvement: territoires et réseaux de migrants albanais en Grèce*. collection 'L'Atelier méditerranéen', Paris: Karthala, Maison méditerranéenne des sciences de l'homme, École française d'Athènes.

Sintès P. 2010b. "Construction des discours d'appartenance en migration: l'exemple des Albanais en Grèce". *Revue ANATOLI*, "L'intégration régionale en Europe du Sud-est", 1(1): 195-213.

Sintès P. 2010c. "Retrouver Rhodes". *Téoros. Revue francophone de recherche en tourisme*. 'Le tourisme des Racines'. Quebec: Presses universitaires du Québec: 37-45.

Sintès P. 2011a. "Manœuvres politiques et discours de l'urgence: la frontière égéenne de la Grèce au cœur d'enjeux nationaux et européens", in N. Ribas-Mateos (ed.), *El Rio Bravo Mediterráneo. Las Regiones Fronterizas en la época de la Globalización*. Barcelona: Bellaterra: 375-390.

Sintès P. 2011b. "A qui Rhodes? Discours d'appropriation et de mise en valeur touristique dans la vieille ville de Rhodes". In *Tourisme et site du patrimoine mondial de l'UNESCO*. Quebec: Université Laval: 1096-1108.

Soja E.W. 2000. *Postmetropolis: Critical Studies of Cities and Regions*. Oxford: Basil Blackwell.

Stahl P.-H. 1977. "Soi-même et les autres, quelques exemples balkaniques". C. Lévi-Strauss (ed.) *L'identité*. Paris: PUF (Quadrige collection): 287-303.

Tarrius A. 2007. *La remontée des sud. Afghans et Marocains en Europe méridonale*. La Tour d'Aigues: Éditions de l'Aube.

Triandafyllidou A. and Veikou M. 2002. "The hierarchy of Greekness, ethnic and national identity considerations in Greek immigration policy". *Ethnicities*, 2(2): 189-208.

Postface

The Border: Political Analyser of a Fluid World

Evelyne Ritaine

In a globalized world in which everything is supposed to circulate, the object "border" provides an analyser for the contradictions that hinder, or reorganize, the fluidity of exchanges (Bauman, 2000, 2002).[1]

The Mediterranean region offers us a laboratory for observing these contradictions as demonstrated in the articles published here. Borders in this region have been tightened in recent years, whether as a result of political/territorial conflicts (Israel and Palestine, Morocco and the Sahrawi Territories, Republic of Cyprus and Northern Cyprus) or as part of policies seeking to control immigration (the European Union's external frontiers in the Mediterranean, Israel's southern border). Often, borders are increasingly "armoured" (*blindées*) along the territory and extended by numerous remote controls (Ritaine, 2009, 2012).[2]

Armoured borders run along the fault lines of economic development and reveal differences in wealth which, in the Mediterranean, are amongst the highest in the world (Milanovic, 2005). Everywhere, they appear to separate the rich from the poor: Israel and Palestine, Spain and Morocco, Italy and the Balkans, the European Union (EU) and the African continent. So it is not surprising that one of the most common ways of circumventing borders is economic: crossing economies, whether legal or illegal, benefit from the wealth gap, enriching people who know how to play the border (entrepreneurs involved in relocation, smugglers and traffickers, all types of 'brokers'), and permitting survival for those who are only able to transgress boundaries (illegal workers).

---

1. *Analyzer*: dispositive or event which reveals the contradictions within an institution (definition drawn from institutional analysis).

2. The term 'armoured' is used here to refer borders that are reinforced by and because of political issues and therefore differ substantially from borders where simple checks are the expression of state sovereignty. 'Armoured borders' (here) implies the deployment of military resources and preconceptions on the suspicious nature of trade and/or crossings and an asymmetric interpretation of the situation (Ritaine, 2009). Reinforcement of territorial borders occurs as part of more substantial remote controls and the use of virtual technologies: these constitute an omnipresent "world-border" for everyone (Cuttitta, 2007).

Reinforcing borders occurs when policies have failed. This is particularly true where contested territorial limits are barricaded, for example, those separating Israel and Palestine: because there is no political settlement, technology must be used to separate the populations, thus creating a series of non-solutions that, over time, contribute to the exacerbation of conflicts. Armouring borders against migrants is, thus, an indication of an absence of migration policies adapted to the new conditions of global order (or disorder). Thus, the EU has created its external border, as a result of free circulation within the Schengen area, without having first established a common approach for its management and shared principles for its immigration policy. Political decisions on borders are taken at the inter-governmental and EU level, but lack conviction. The EU does not have a decision-making body for these policies, similar to that of the United States Federal Government. For the Mediterranean Basin, it is the Southern European countries, in the first instance, that find themselves responsible for controlling these borders and for managing relations with the EU's neighbours. European policy on the shared management of frontiers has specific requirements and practices that must be incorporated and this makes the search for community-wide approach to these questions particularly difficult. Finally, local authorities (regional or municipal) have to handle emergency situations in the most easily accessible sites: islands, enclaves, and coasts. Armouring the borders is perhaps not so much the result of unequivocal decision-making, but of an accumulation of local decisions taken by politicians in response to a series of emergencies, which have to be explained and integrated after the event. Yet in Europe, the border is not merely a collection of national frontiers for peripheral states, it is also a shared border imbued "with common state-like attributes" (Snyder, 2000: 225). This complex and political state of uncertainty lies behind the many ways that European principles, with regard to the management of shared borders, have wavered.: it explains the difficulties in sharing the social responsibility for managing borders and lawless areas that prevail in the handling of clandestine migrants.

Nevertheless, there is a strategy behind this political failure. The armouring of border represents a kind of "gallant last-stand" to the Westphalian State which has been deprived from most of its capacity for intervention (Brown, 2009): thus Southern European States seek to reaffirm their sovereignty by attempting to maintain political control over access to their territory. To achieve this, they "govern by anxiety" which encourages exclusionary populism and the refusal of global interdependence (Bigo, 1998; Betz, 2001; Ritaine, 2005). Based on the reaffirmation of a territorial political community, these political positions explain why people feel insecure about dangers from 'out there' (globalization of trade, supra-national institutions, presence of mafias, terrorists, clandestine immigrants, delinquents, etc.). As part of defensive strategies, they advocate tighter controls for ensuring public and personal security inside the country (the security argument), protection of the local or national cultural identity (the identity argument, which often goes hand-in-hand with the fantasy of demographic dilution), and restriction on social benefits to the 'native-born' population (the exclusive social protection argument).

Exclusory populism is usually expressed through representations of 'neighbour as threat', 'contiguity as risk', 'frontier as the frontline between us and them', 'barricaded borders as protection'. In particular, they were used by the governments led by Aznar (Spain) and Berlusconi (Italy) and as an argument for Greece's on-going refusal to establish a coherent policy on migration. They have thus reinforced the tendency to focus principally on security issues in developing policy for managing the EU's frontier in the Mediterranean (Ritaine, 2012). context of siege mentality, barricading the Mediterranean has become a complex geopolitical issue, as can be seen in the contradictory positions taken by European Member States with regard to the 'Arab Spring' in 2011: on the one hand, support for democratic movements while, on the other, horrified reactions to the arrival of (modest) numbers of migrants on European beaches.

Similarly, Israel's introduction of multiple systems for controlling the Palestinian population goes way beyond a simple national policy on security. For this reason, it has resorted to tactics, based on the rejection of the Other, in order to maintain political consensus and, possibly, to shore up political coalitions. Above all, this makes it possible to control Palestinians by a complex collection of physical and regulatory constraints that prevents their normal social and economic activities and, consequently, hinders all attempts to establish a Palestinian state (Weizman, 2007; Ophir *et al.*, 2009).

Finally, strategies of political exclusion have a dramatic and symbolic impact: they maintain the dichotomy between a domestic community and an external Other, between the secure space and the external danger. It also creates a shared identity on one side, and a shared stigmatisation on the other. But, in a globalized world where there is no longer an 'outside',[3] the use of such symbols of toughness only serve to maintain the potential for conflict between 'enemy' groups. These forms of symbolic opposition contribute to repeated failures of trans-Mediterranean partnerships (Union of the Mediterranean, Union for the Mediterranean); since these partnerships have become bogged down in

3 Expression coined by Ceyhan (2001).

asymmetric relations between the two sides of the Mediterranean, with Europe seeking to export its "good practices":

> Euro-Mediterranean policies suffer from these horizons in the sense that the Other is Evil. But mimetic imaginaries also constitute reference systems in establishing security policies in the region. It is these "imponderables" that reactivate boundaries and the necessity of defending them. [...] From now on, Euro-Mediterranean policies are influenced by questions of security, which efforts to render situations more social and accommodating are less important than schemes for police and customs controls (Hastings, 2011: 32).

It is this fantastic dichotomy, as much as the concrete obstacles, that populations "on the wrong side" must confront. Nevertheless, social dynamics are obstinate and they reinvest this cleavage with all the force of a population's ability to adapt. People adapt to controls and closures, they transgress imposed limits and take advantage of new border configurations. Once inside the country, the presence of internal frontiers encourage migrants to resort to ruses and local knowledge in order to find their way in urban areas, avoid police controls, and develop strategies that will ensure their invisibility. Similarly, historical time, based on long-established practices and habitual interactions, is not easily forgotten: it continues to influence daily activities and collective memory and this is all the more true in that they are sustained by well-established groups of immigrant origin. Finally, mobilisation militancy, on both sides of the border, contributes to the complexity of *rapports de force*: in addition to state authorities, civil society is increasingly involved in questions relating to borders by protests in favour of the right to free circulation and calls for respect of human rights at the border.[4]

Also, permanently swamped by these social dynamics, today's hyper-frontier is continually reproduced and reinforced, indeed more barricaded by recourse to military technologies as borders become increasingly 'smart' through the introduction of new surveillance technologies. Everywhere, we see the situation on a collision course with more levels of surveillance introduced each time a new chink appears in the barricade. Everywhere, by inhibiting some people's ability to travel, the situation becomes a self-fulfilling prophecy: more barricades lead to more people trying to cross the border; more discrimination on control points leads to more disorder; more categorization leads to more inequality between different social categories. At the frontiers themselves, as political and administrative procedures become increasingly arbitrary,

---

4 In the Mediterranean, NGOs protest against the installation of retention centres as a tool for managing illegal migrants and against the risks of death that arise from the fortification of borders (it is estimated that as many as 12,000 people have been lost at sea in 10 years) (Agier, 2010).

so people invent more ways to circumvent the obstacles: in human society, collective intelligence is often expressed in people's ability to detour.

References

Andreas P. and T. Snyder (eds.) 2000. *The Wall around the West: State Borders and Immigration Controls in North America and Europe*, Lanham (MD): Rowman & Littlefield.

Agier M. 2010. "Couloirs d'exil. Un paysage global de camps", *Métropolitiques*, http://www.metropolitiques.eu/Couloirs-d-exil-Un-paysage-global.html; http://fortresseurope.blogspot.fr/p/la-strage-negata-17317-morti-ai-confini.html.

Bauman Z. 2000. *Liquid Modernity.* Cambridge/Oxford: Polity Press/ Blackwell Publishers.

Bauman Z. 2002. *Society Under Siege.* Oxford: Blackwell Publishers.

Betz H.-G., 1998. *The New Politics of the Right: Neo-Populist Parties and Movements in Established Democracies*, Macmillan,

Bigo D. 1998. "Sécurité et immigration: vers une gouvernementalité par l'inquiétude?" *Cultures et Conflits*, 31/32:13-38. http://conflits.revues.org/539.

Brown W. 2009. *Murs. Les murs de séparation et le déclin de la souveraineté étatique.* Paris: Les Prairies ordinaires.

Ceyhan A. 2001. "La fin de l'en-dehors: les nouvelles constructions discursives de l'ennemi intérieur en Californie". *Cultures & Conflits* 43, http://conflits.revues. org/569.

Cuttitta P. 2007. "Le monde-frontière. Le contrôle de l'immigration dans l'espace globalisé". *Cultures et Conflits* 68 Circulation et archipels de l'exception, http://conflits.revues.org/index5593.html.

Hastings M. 2011. "Le 'méditerranéisme': un nouvel orientalisme?", in Matuszewicz R. and L. Beurdeley (eds.), *Regards croisés sur les sécurités dans les relations euro-méditerranéennes*. Paris: L'Harmattan: 25-36.

Matuszewicz R. and L. Beurdeley (eds.) 2011. *Regards croisés sur les sécurités dans les relations euro-méditerranéennes*. Paris: L'Harmattan.

Milanovic B. 2005. *Worlds Apart. Measuring International and Global Inequality*, Princeton, Oxford, Princeton University Press.

Ophir A., M. Givoni, and S. Hanafi 2009. *The Power of Inclusive Exclusion. Anatomy of Israeli Rule in the Occupied Palestinian Territories*, New York, Zone Books.

Ritaine E. (ed.) 2005. *L'Europe du Sud face à l'immigration. Politique de l'Étranger*. Paris: PUF, Sociologie d'aujourd'hui.

Ritaine E. 2009. "La barrière et le checkpoint: mise en politique de l'asymétrie". *Cultures et Conflits* 73 Frontières, marquages et disputes: 13-33. http://www.cairn.info/revue-cultures-et-conflits-2009-1-page-15.htm.

Ritaine E. 2012. *La fabrique politique d'une frontière européenne en Méditerranéen? Le "jeu du mistigri" entre États et Union.* Les Études du CERI, 186 http://www.sciencespo.fr/ceri/en/papier/etude.

Snyder T. 2000. "Conclusion: The Wall around the West", in Andreas P. and T. Snyder (eds.), *The Wall around the West: State Borders and Immigration Controls in North America and Europe* Lanham (MD): Rowman and Littlefield: 219-227.

Weizman E. 2007. *Hollow Land. Israel's Architecture of Occupation*. New York, London: Verso.

Shorts Biographies of the Authors

Ilsen About is an historian, post-doctoral researcher at the Institute of Mediterranean and Comparative Ethnology (IDEMEC) in Aix-en-Provence. At the moment, he is involved in a postdoctoral project comparing the different politics and policing of the "Gypsies" in Western Europe, including Italy, France, Switzerland, Belgium and Germany, between 1880 and 1930. Another postdoctoral project is dedicated to the institutionalisation of a French police department specialised in the identification, control and surveillance of the foreigners during the interwar period. He is the author with Vincent Denis of *Histoire de l'identification des personnes* (La Découverte, 2010).

Lisa Anteby-Yemini is an anthropologist, Researcher at the CNRS (French National Center for Research) and member of the Institute of Mediterranean and Comparative Ethnology (IDEMEC) in Aix-en-Provence. She has researched extensively the Ethiopian Jewish community in Israel and is author of "Les Paradoxes du paradis: les juifs éthiopiens en Israël" (2004, Éditions du CNRS, Paris). She is currently working on non-Jewish migration to Israel and particularly on African asylum-seekers and has published articles on the subject. She is also editing a book on Women in Judaism and Islam resulting from a three-year European research program she led on this topic.

Virginie Baby-Collin is associate professor in the department of geography and planning, Aix Marseille University since 2004, Telemme research Center, MMSH (Maison méditerranéenne des sciences de l'homme). Since 2010, she is a Visiting scholar at Columbia University (New York City). Her research in urban comparative geography has expanded to the study of Latino American immigration in Southern Europe (Spain and France), and more recently in the New York metropolitan area. Her main interests are international migration issues, transnational ways of life, territories and identities, fragmentation and cosmopolitanism in metropolitan contexts. Among her recent publications, the coedition of various books: *La famille transnationale dans tous ses états* (2011, *Autrepart*, Presses de sciences Po), *Cuando México enfrenta la globalización. La zona metropolitana de Monterrey* (2010, UANL, Plaza y Valdez editores), *Migrations et territoires de la mobilité dans l'espace méditerranéen* (2009, *Méditerranée*), *Migrants des Suds* (2009, IRD-PUM).

Assaf Dahdah is a geographer in Aix-Marseille University in TELEMMe laboratory. He is currently pursuing a PhD on non-Arab migrants in Beirut, Lebanon. His research focuses on the visibility of migrant workers' presence in the city and situations of co-presence more specifically in marginal spaces at the edges of the capital with the resulting urban transformations. Based on urban and migration studies, his work and approach try to understand the place of "otherness" in a Mediterranean metropolis like Beirut.

Karima Dirèche is a historian, Researcher at the CNRS (French National Center for Research) and member of Telemme research Center in Aix-en-Provence and of Jacques Berque Center for human and social sciences in Rabat. She is a specialist of social history of the colonial and post-colonial Algeria. Among her publications: *Femmes, familles et droit au Maghreb* (CNRS, 2007).

Heidrun Friese is currently visiting fellow at the EUI and working on a research project on the 'Limits of Hospitality' (funded by the German Research Fund, DFG). Her publications include Special issue on Transborders. *European Journal of Social Theory* (co-edited with S. Mezzadra), 2010; 'Europe's otherness. Cosmopolitism and the construction of cultural unities'. In: Gerard Delanty (ed.), *Europe and Asia Beyond East and West: Towards a New Cosmopolitanism*, London, Routledge 2006*; 'Cultural identities*'. In: Gerard Delanty (ed.), *Handbook of Contemporary European Social Theory*. London: Routledge, 2006; *Europa, costituzione e movimenti sociali.* (co-edited), Rome, manifestolibri, 2003; *Identities. Time, Boundaries and Difference* (solely edited), Oxford, Berghahn 2002; *Europa politica. Ragioni di una necessità* (co-edited), Rome, manifestolibri, 2002; *The Moment. Time and Rupture in Modern Thought* (solely edited), Liverpool, Liverpool University Press, 2001.

Arnaud Garcette is a PhD candidate at Institut d'Études et de Recherche sur le Monde Arabe et Musulman (IREMAM, Aix-Marseille University). He studies the readjustment of the Palestinian Olive sector since 1993 from a socio-economic perspective. He has carried out a two years field work in Israel Palestine in 2010-2011.

Fabrice Jesné is is associate professor in contemporary history at University of Nantes and former member of French Institut in Roma (École française de Rome). His main interests are the relationships between Italy and Balkan but also the consular network in mediterranean in 19th century.

Nick Mai is Reader in Migration Studies at ISET, the Institute for the Study of European Transformations of London Metropolitan University. Nick researched several aspects of the Albanian post-communist transformation, with a particular focus on the role played by Italian media in the eliciting of migration and in stigmatising Albanian migrants in Italy. In the last 6 years, Nick researched the mobility of minors and young migrants from Eastern Europe (Albania, Romania) and North Africa (Morocco, Tunisia) into and within the EU, their strategies of survival and the associated risks and opportunities, including issues of exploitation and the engagement in illegal activities. His current research focuses specifically on the global sex industry as a space within which migrants (and non migrants) both challenge and reproduce established intersections between social mobility, gender and sexuality.

Sylvie Mazzella is a sociologist, Researcher at the CNRS (French National Center for Research) and member of Mediterreanean Institut of Sociology (LAMES) in Aix-en-Provence. Her researches concern migrations South-North, South-South from North-Africa. She edited *La mondialisation étudiante. Le Maghreb entre Nord et Sud* (Karthala, 2009)

Stéphane Mourlane is associate professor in contemporary history at the Aix Marseille University, member of Telemme research Center, MMSH (Maison méditerranéenne des sciences de l'homme) in Aix-en-Provence. His main interest is the cultural history of international relationships in Mediterranean in the second part of the XXth century. His research concern in particulary Modern Italy: migration, cultural Institutes and sport. Among his recent publications, *Nice cosmopolite, 1860-1960* (Autrement, 2010), *Atlas de l'Italie contemporaine* (Autrement, 2011) and *Empreintes italiennes. Marseille et sa région* (Lieux-Dits, 2013). He is also editing a book with Céline Regnard: *Les batailles de Marseille. Migrations, violences et conflits* (Presses universitaires de Provence, 2013).

Cédric Parizot is an anthropologist, Researcher at the CNRS (French National Center for Research) and member of l'Institut d'Etudes et de Recherche sur le Monde Arabe et Musulman (IREMAM, Aix en Provence). His Research focuses on mobilities and frontiers in Israeli-Palestinian space. He edited with Stéphanie Latte Addallah *À l'ombre du mur: Israéliens et Palestiniens entre separation et occupation* (Actes Sud, 2001).

Laurence Pillant is a geographer in Aix-Marseille University in TELEMMe laboratory. She is currently pursuing a PhD on the control of irregular migration in Greece. Her research focuses on the territories and the places of this control, and the evolutions of the whole dispositive in the recent years.

Céline Regnard is associate professor in contemporary history at the Aix Marseille University, member of Telemme research Center, MMSH (Maison méditerranéenne des sciences de l'homme) in Aix-en-Provence.

Her research focuses on the links between migrations and violence in XIXth and XXth century. She is the author of *Marseille la violente. Criminalité, industrialisation et société 1851-1914*. (Presses universitaires de Rennes, 2009) and *Empreintes italiennes. Marseille et sa région* (Lieux-Dits, 2013). She is also editing a book with Stéphane Mourlane: *Les batailles de Marseille. Migrations, violences et conflits* (Presses universitaires de Provence, 2013).

Sandro Rinauro is associate professor in contemporary history at Milano University. His research concern italian emigration. He is the author of *Il cammino della speranza. L'emigrazione clandestina degli italiani nel secondo dopoguerra* (Einaudi, 2009).

Evelyne Ritaine is Researcher at the National Fondation of Politic Science (FNSP), member of Emile Durkheim center, associated at the Centre for International Studies and Research (CERI-SciencesPo Paris). She is also member of the network Association for Borderlands Studies. Her research investigates contemporary walls (armour-plated borders, contested ceasefire lines, urban separations, gated communities) as comparable gestures reflecting a process of reinforcement of a territorialised limit aimed at controlling the circulation of persons. She edited *Politique de l'Etranger: l'Europe du sud face à l'immigration* (PUF, 2005).

Pierre Sintès is associate professor in the department of geography and planning, Aix Marseille University since 2006, Telemme research Center, MMSH (Maison méditerranéenne des sciences de l'homme). His main interests are international migration issues, transnational ways of life, territories and identities, fragmentation and cosmopolitanism in the Greek and Balkan societies. Among his recent publications: *Atlas géopolitique des Balkans* (2011, Autrement), *Géographie des conflits* (2011, Bréal), *La raison du mouvement: territoires et réseaux de migrants albanais en Grèce* (2010, Karthala, Maison méditerranéenne des sciences de l'homme), *Migrations et territoires de la mobilité dans l'espace méditerranéen* (2009, *Méditerranée*); and the coedition of various books: *Migrations et territoires de la mobilité dans l'espace méditerranéen* (2009, *Méditerranée*), *Nommer et classer dans les Balkans* (2008, École française d'Athènes).

Regional Integration and Social Cohesion (RISC)

Regional Integration and Social Cohesion (RISC) is an interdisciplinary and multi-lingual (English, French, German, Spanish, Italian and Portuguese) series that examines the human and environmental impacts of regional integration, defined legally/politically (supranationalism), economically (globalisation), socially (identity) and geographically (borders). The series aims to link global analysis of issues associated with social cohesion, such as market shifts, immigration, environmental risk and the deterioration of natural resources, human rights, violence and security, etc., with national and sub-national studies that focus on political decision-making, the non-governmental sector and social participation in public debates, economic transformations, social marginalisation, identity, etc. Thus, the series aims to combine the international, national and sub-national arenas of politics in thematically-oriented research. While proposed case studies are welcome, the works presented in this series from all social science disciplines, are predominantly based on the comparative method. Studies that compare across geographic regions, defined continentally, are of particular interest. In addition to the scientific focus described above, this series aims to connect theoretical analysis of questions related to social cohesion with policy-based research. In doing so, it examines the role of political actors at different levels of regional integration processes and it studies citizen responses to changing opportunity structures in the economic, social and political spheres. Thus, the series attempts to shed light on contemporary shifts in the uses and types of power in policy-making processes. The issue of changes in how policies are being made is linked to: "Who affects policy-making?" and "What impact do policies have in social and economic arenas?"

Juan Carlos Velez Rendón, *Universidad de Antioquia, Colombia*
Lucelly Villegas Villegas, *Universidad de Antioquia, Colombia*
Casey Walsh, *University of California, Santa Barbara, USA*

Published Books

N° 13 – Lisa ANTEBY-YEMINI, Virginie BABY-COLLIN, Sylvie MAZZELLA, Stéphane MOURLANE, Cédric PARIZOT, Céline REGNARD & Pierre SINTÈS (eds.), *Borders, Mobilities and Migrations. Perspectives from the Mediterranean, 19-21st Century*, 2014.

N° 12 – Candice MOORE, *Regional Integration and Social Cohesion: Perspectives from the Developing World. RISC 2011*, 2013

N° 11 – Claudia PUERTA SILVA, *Stratégies et politiques de reconnaissance et d'identité. Les Indiens wayuu et le projet minier du Cerrejón en Colombie*, 2013

No.10 – Carmen MAGANDA & Olivier PETIT (eds.), *Strategic Natural Resource Governance. Contemporary Environmental Perspectives / La gouvernance des ressources naturelles stratégiques. Perspectives contemporaines dans le domaine de l'environnement*, 2012

No.9 – Peter GILLES, Harlan KOFF, Carmen MAGANDA & Christian SCHULZ (eds.), *Theorizing Borders Through Analyses of Power Relationships*, 2013

No.8 – Julia AFFOLDERBACH, Travis DU BRY, Olga L. GONZALEZ & Constanza PARRA (eds.), *Reinforcing Governance. Perspectives on Development, Poverty and Global Crises. RISC 2010*, 2012

N° 7 – Oana AILENEI, *Le rôle de l'économie sociale dans les dynamiques socioéconomiques locales. Des concepts aux initiatives d'innovation sociale à l'échelle des quartiers en Europe et à Roubaix, France*, 2011

No.6 – Claudia PUERTA SILVA y Juan Carlos VÉLEZ RENDÓN (eds.), *Fronteras y reconfiguraciones regionales. RISC 2009*, 2011

No.5 – Felipe J. HEVIA, *Poder y ciudadanía en el combate a la pobreza. El caso de Progresa/Oportunidades de México*, 2011

No.4 – Carmen MAGANDA & Harlan KOFF (eds.), *Perspectivas comparativas del Liderazgo. RISC 2008 / Comparative Perspectives on Leadership. RISC 2008*, 2009

No.3 – Harlan KOFF (ed.), *Social Cohesion in Europe and the Americas. Power, Time and Space / Cohesión social en Europa y las Américas. Poder, tiempo y espacio*, 2009

No.2 – Adriana GONZÁLEZ GIL (ed.), *Lugares, procesos y migrantes. Aspectos de la migración colombiana*, 2009

No.1 – Harlan KOFF, *Fortress Europe or a Europe of Fortresses? The Integration of Migrants in Western Europe*, 2008

www.peterlang.com